A Sonoran Desert Scrapbook

Some Desert Plants of Kino Bay and Vicinity

WILLIAM J. LITTLE

ARIZONA-SONORA DESERT MUSEUM PRESS

Second Edition published by Arizona-Sonora Desert Museum Press 2022
© 2022 Arizona-Sonora Desert Museum
All Rights Reserved.

Author photographs © 2011 William J. Little
All photographs, unless otherwise noted, are by the author.

Grateful acknowledement is made to the following for permission to reprint
previously published material: Christopher L. Helms for map of The Four Great
Deserts of North America (page 3) and map of The Seven Divisions of the Sonoran
Desert (page 7). University of Arizona Press for photo (page 11) by Richard Felger in
People of the Desert and Sea: Ethnobotany of the Seri Indians by Richard S. Felger
and Mary Beck Moser; 1985 The Arizona Board of Regents. Bert Wilson,
www.laspilitas.com, for use of photo (page 109) of Desert Lavender.

Book design and production by Carolyn Kinsman.
Printed by IngramSpark in the United States of America.

COVER PHOTO: Cardón cactus (Pachycereus pringlei) along the highway entering
Kino Bay.

Originally published by Dog Ear Publishing in the United States of America in 2012.

ISBN: 979-8-218-02660-8
1. Plants, Botany, Ethnobotany
2. Sonoran Desert
3. Seri Indians
4. Kino Bay, Bahía de Kino, Sonora, Mexico
5. Arizona

ARIZONA-SONORA
DESERT
MUSEUM

Arizona-Sonora Desert Museum
2021 N. Kinney Road
Tucson, AZ 85743

desertmuseum.org
asdmpress@desertmuseum.org

ABUNDANCE

The abundance of the desert
 can be a surprise
even for those
who call it home.

The variety of its species,
the range of its landscapes,
the vastness of its space
all come together
to capture your heart.

No matter the season,
or the region,
you know you're
someplace special
when you're in the
Sonoran Desert.

From a brochure by the
Arizona-Sonora Desert Museum

Tucson, Arizona

Estados Unidos de Améri

A R I Z O N A

Estado de Sonora

Golfo de California

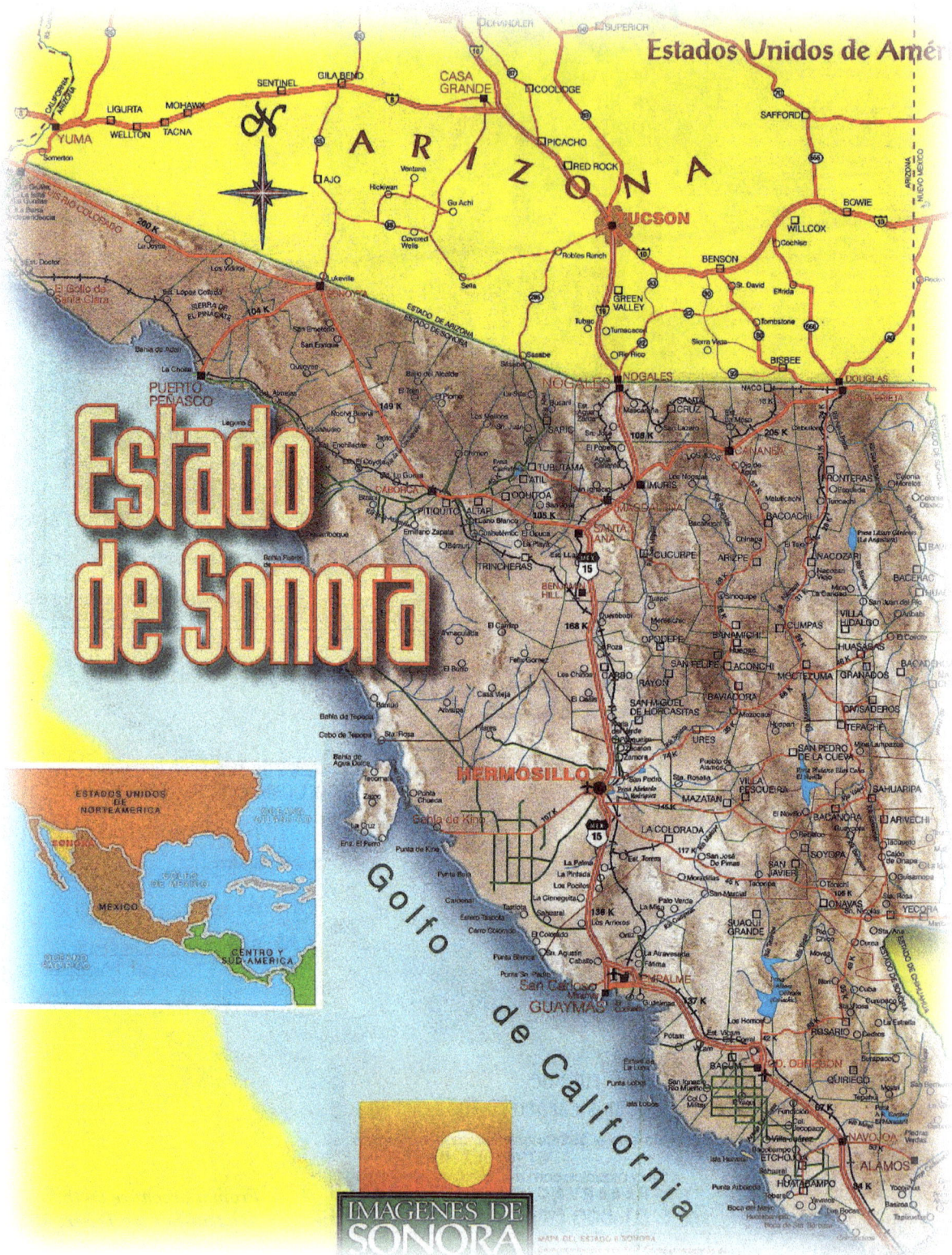

ESTADOS UNIDOS DE NORTEAMERICA

MEXICO

CENTRO Y SUD-AMERICA

IMAGENES DE SONORA

Contents

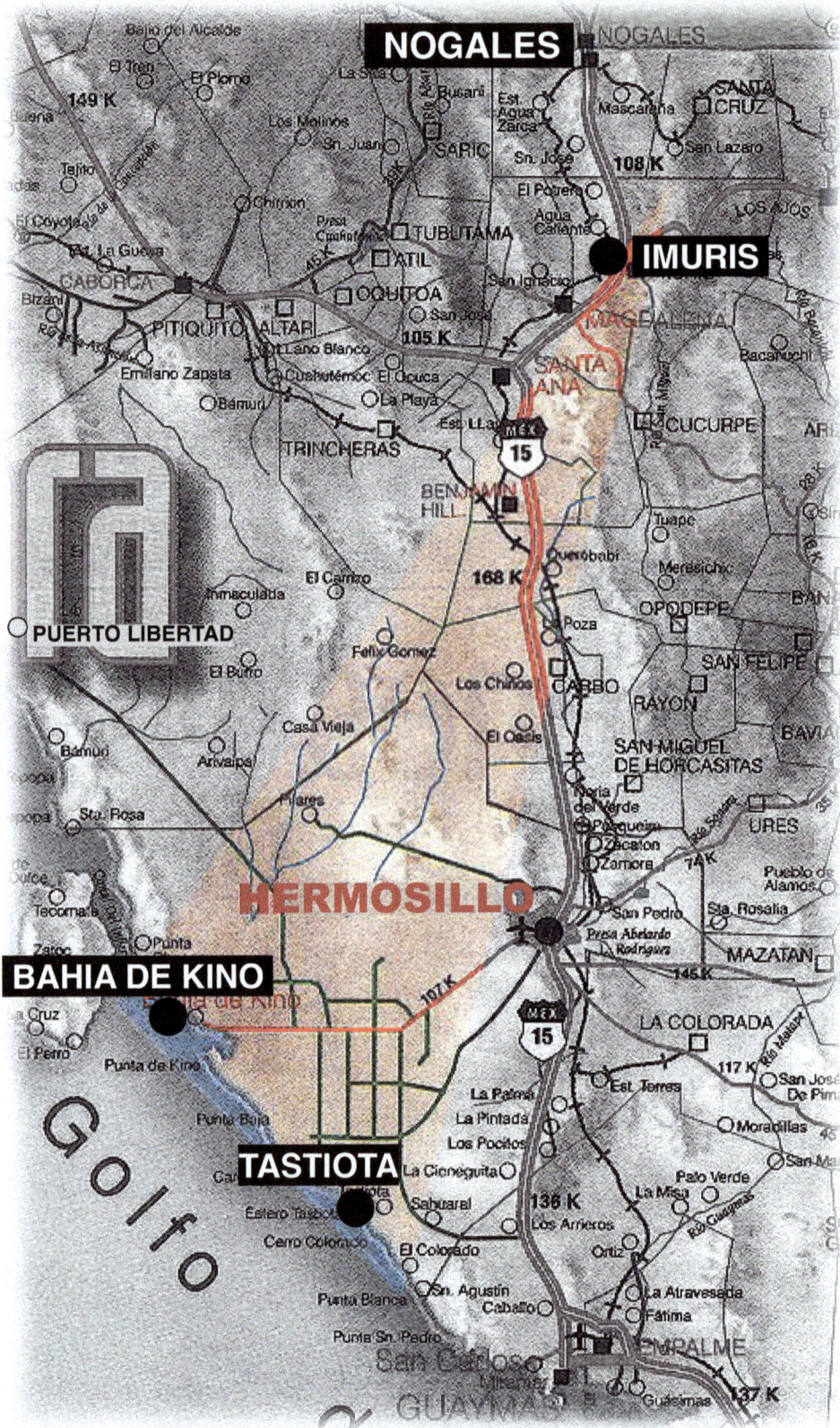

Introduction and Acknowledgments

Mexico has the fourth richest biodiversity of all countries in the world (*Nature Conservancy* magazine, Vol. 2, No. 4). Sharing in this biodiversity is a large part of the Sonoran Desert contained within the Baja peninsula and the State of Sonora. This book introduces the Sonoran Desert and takes the reader on a tour of five areas in and around Kino Bay, Sonora to see a variety of plants and learn something of their ecology and ethnobotany.

Visulize the area inside a triangle formed by Imuris, Tastiota and Kino Bay, Sonora. Here you'll find a variety of plants including the beautiful blue-flowered Guayacán of "Seri blue dye" fame; the weird Boojum tree; the graceful White-bark Acacia, endemic to a narrow strip along the coast between Kino Bay and Guaymas; and six species of columnar cacti including the massive Cardón, the world's largest cactus.

The plants in this book are organized into five areas, although most can be seen in more than one area.

Area A. Roadside plants from Nogales to Kino Bay

Area B. Plants of the dunes, estuaries and coastal wetlands

Area C. Plants of Kino Bay and vicinity

Area D. Plants of the Tastiota area south of Kino Bay

Area E. Plants of the Puerto Libertád area north of Kino Bay

The chart on page 148 shows the relative abundance of each plant within each of the five areas.

Most plants can be seen from roadsides or a short walk. The more adventurous will be rewarded!

Material for the book was gathered from many sources including books cited in Selected References, page 147, and the field experience of the author.

Appreciation is extended to the staff of the Tucson Botanical Garden, John Wiens of the Arizona-Sonora Desert Museum, Tucson, and Phil Jenkins of the University of Arizona, Tucson, for assistance in the identification or verification of many of the plants presented here. Desert maps are from *Sonoran Deserts: The Story Behind the Scenery,* courtesy of Christopher Helms. The Sonora state map is from *Hermosillo and Bahía de Kino,* a publication for tourists by Imagenes de Sonora, Hermosillo, Sonora, 2004. Much of the ethnobotany information of the Seri Indians is from *People of the Desert and Sea, An Ethnobotany of the Seri Indians* by Richard Stephen Felger and Mary Beck Moser, ©1985, published by The University of Arizona Press, courtesy of the Arizona Board of Regents. Photographs are by the author unless otherwise noted.

I greatly appreciate the contribution of John Wiens who did a peer review of the manuscript and made corrections and suggestions that greatly improved the botanical aspects of the book.

A special thanks to Carolyn Kinsman who edited the manuscript and prepared it for publication. Thanks to Corinne Herpel for editing of the original draft, to Perry R. Wilkes for editing and proofing of the final version, and to Luis Healy for years of assistance in making the Spanish version possible. Residual editing errors are mine.

I am grateful to my wife, Mary, who assisted with preparation of the first book, and shared my love for the Sonoran Desert.

— *William J. Little*

Luis Alberto Healy Loera
(1959-2021)

This book is dedicated to the memory of Luis Alberto Healy Loera, my very good friend who always gave generously of his time and talents to so many others. He was one of the finest people I ever met. — *William J. Little*

"Everything will be
okay in the end.
If it's not okay
it's not the end."
– *Luis Alberto Healy Loera*

An Introduction to the Sonoran Desert

The foundation for much of what we know of the plants and ecology of the Sonoran Desert came from pioneering work by Forrest Shreve and his colleagues at the Carnegie Desert Botanical Laboratory on Tumamoc Hill just west of Tucson. The laboratory was created by the Carnegie Institution of Washington D.C. in 1902 "... *for the purpose of ascertaining how plants perform their functions under extraordinary conditions existing in desert.*" It was the first research institution in the world to be devoted entirely to desert problems, and the work stimulated other desert research throughout the world. *Carnegiea*, the genus name for Saguaro, was given to this famous cactus in recognition of the support to desert plant science by the Carnegie family.

Although a few wandering botanists had found and named most of the plants in the Sonoran Desert prior to 1902, Shreve and his associates were the first to study these plants and their ecology in detail. The laboratory closed in 1940, and the property with its stone buildings is now owned by the University of Arizona.

Many scientists from Mexico and the United States continue to broaden the scope of our knowledge of the plants, wildlife, and ecology of the Sonoran Desert.

The Four Great Deserts

Forrest Shreve delineated the North American deserts into four "Great Deserts" based on their distinctive vegetation, which is a product of their unique climates and soils. They are the Great Basin, Mohave, Chihuahuan, and Sonoran Deserts. Lesser deserts are included within the Great Deserts. An example is the Gran Desierto at the mouth of the Colorado River which is part of the Sonoran Desert.

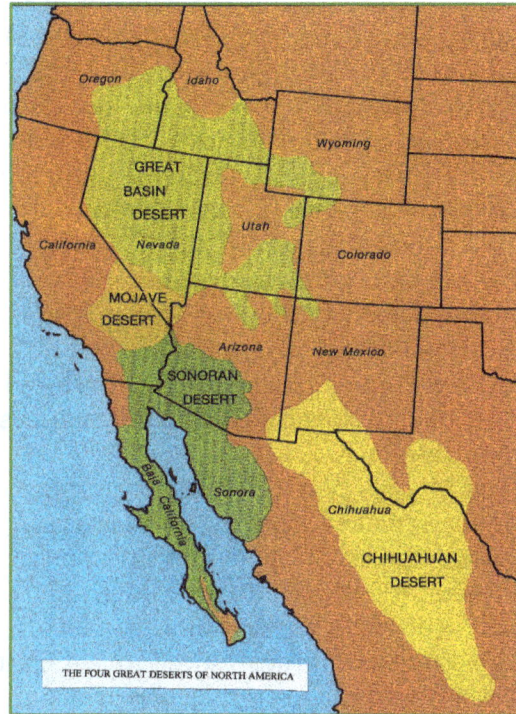

THE FOUR GREAT DESERTS OF NORTH AMERICA

Many things interact to differentiate the four Great Deserts, but temperature regimes and season(s) of precipitation have the greatest influence, as will be seen in the following descriptions.

Great Basin Desert

The Great Basin Desert is located in parts of Idaho, Oregon, Wyoming, Colorado, Utah, Nevada, Arizona, and New Mexico. It is known as the Cold Desert, as winter temperatures often drop below freezing and may reach -25 degrees or lower. About 60% of precipitation comes in winter. Plants have adapted to a short growing season and major temperature extremes. For most plants, there is a single flowering period which occurs in the spring and early summer. A few plants such as sagebrush and rabbitbrush flower in late summer or early fall. The typical vegetative aspect is sagebrush-grass or saltbush-grass.

Mojave Desert

The Mohave Desert is located in southeastern California, southern Nevada, and the northwestern tip of Arizona. It is the most arid of the four Great Deserts. A temperature of 134 degrees F was recorded in Death Valley in 1913, second only to the world record of 136 degrees F recorded in the Sahara Desert at Al Aziziyah, Libya in the 1930s. Winters are cool. Annual precipitation ranges from 2 to 5 inches, falling primarily in winter. There is a single flowering period in the late winter-spring. Vegetation is typically sparse. High alkalinity in the surface horizons (related, in part, to sparse rainfall) further reduces species diversity.

The Mojave has the smallest number of plant species of the four Great Deserts. Trees are absent over most of this desert. The lowest point is Badwater Basin in Death Valley at 282 feet below sea level. The Joshua tree, largest yucca in the United States, grows only at higher elevations in the Mojave. The typical plant aspect is a desert of short shrubs and winter annuals.

Chihuahuan Desert

This is a high desert located mostly in northern Mexico where it is bordered on the east by the Sierra Madre Oriental and on the west by the Sierra Madre Occidental. Between these two mountain ranges is the Altiplano, or high plains of northern Mexico with an average altitude of 4,000 feet. Portions of the desert extend into Texas, New Mexico, and Arizona.

Precipitation ranges from 3 to 12 inches with 65 to 80% occurring as summer rain from mid-June to mid-September. Moisture is scant the remainder of the year. Summer temperatures average 10 to 20 degrees cooler than the Sonoran Desert. Freezing temperatures are common in winter. Summer rain and cold winter temperatures result in a single summer-fall flowering period. Leaf succulents, such as agaves, and semi-succulents, such as yuccas, are common. The general plant aspect is a large diversity of shrubs, small cacti, yuccas, and agaves. Trees are rare. Mesquite usually only attains shrub-size. Chihuahua is a Tarahumara Indian word meaning "place of workshop."

Some locations within the Chihuahuan Desert are: Socorro, Truth or Consequences, Lordsburg, Las Cruces, and Carlsbad, New Mexico; Douglas, Arizona; El Paso, Pecos, and Presidio, Texas; Torreon in Coahuila, Mexico; and Monterey in Nuevo Leon, Mexico.

Sonoran Desert

The Sonoran Desert is located in southern Arizona, southeastern California, Baja California, and the state of Sonora, Mexico. It is a low, hot desert where freezing temperatures are uncommon and of short duration. Most precipitation falls during two distinct periods: mid-June to mid-September (the monsoons) and December through March. Summer rainfall amounts to about 65% of the yearly total on the eastern side of the desert and declines to about 40% or less on the western side. The ratio of summer-to-winter rain is the determining factor for the distribution of many species of plants in the Sonoran Desert. Some species, such as Guayacán *(Guaiacum coulteri),* are summer rain-dependent and grow only where there is a higher ratio of summer-to-winter rain. Some species, such as Smoke tree *(Psorothamnus spinosus),* are winter rain-dependent and grow primarily where there is a higher ratio of winter-to-summer rain. Other species, such as Brittlebush *(Encelia farinosa),* show no preference and are widely distributed. The ratio at Kino Bay is about 50-50.

The bi-seasonal rainfall, unique to the Sonoran Desert, produces two distinct flowering periods – summer and winter. Some species bloom once during the year, some twice, and some bloom any time following sufficient rain.

The Sonoran Desert has the warmest winters and largest number of plant species of the four Great Deserts. It is called an "arboreal desert" because of the large number of tree species, including columnar cacti such as Saguaro. Cold temperatures, more than aridity, limits plant diversity (number of species) in the other three Great Deserts.

Creation of the Sonoran Desert

The plant community you see today began forming about 9,000 years ago following the last Ice Age. During the Ice Ages, the weather was cooler and wetter and plants requiring more moisture, like pinion, juniper, and oak, grew over much of the area of the present desert. During interglacial periods, the climate warmed and species such as Brittlebush and Saguaro moved in from more tropical thornscrub forests to the south, only to disappear with the next glacial period. As the Sonoran Desert gradually became hotter and drier during the last 9,000 years, plants with higher water requirements, such as the pinion, juniper, and oak, were eliminated, and more adaptive plants moved in. Many of the plants and animals here today had their origin from ancestors in tropical thornscrub to the south.

Plant Diversity

According to the Arizona-Sonora Desert Museum (Tucson), there are an estimated 2,500 species of plants in the Sonoran Desert. This desert has more species of plants, more types of cacti, and more species of animals than any of the other Great Deserts. It also has the greatest vertical diversity or variation in plant heights. This biological diversity makes it the most biologically rich, colorful, and interesting of all the Great Deserts.

In general, the more arid habitats have a greater proportion of annual plant species. Half of all Sonoran Desert plants are annuals. In the driest parts of the Sonoran Desert, such as the sandy flats near Yuma, 90% of the plant species are annuals. This large variety of annuals produces a riot of color following winters of abundant rainfall.

Precipitation

Annual precipitation in the Sonoran Desert ranges from 14 inches at Roosevelt Lake Ranger Station (Arizona) in the desert's northeast corner to about 1.5 inches in northeastern Baja California near the mouth of the Colorado River. Some rainfall averages in inches are: Santa Ana – 12, Tucson – 11, Hermosillo – 13.5, Guaymas – 7.5, and Puerto Libertád – 4.25. There is no official weather station at Kino Bay, but rainfall is estimated at 5.5 inches per year. Annual averages are highly variable throughout the Sonoran Desert. Some weather stations on the Gulf side of northeastern Baja, California recorded no precipitation for six consecutive years, whereas Kino Bay received an estimated 10 inches during one tropical storm.

Bi-Seasonal Rainfall

The Sonoran Desert has two distinct rainfall periods – summer and winter. In summer, heating at the equator strengthens the Bermuda High, a zone of high pressure in the Atlantic Ocean. This intensifies the moisture-laden tropical circulation systems and pushes them westward across Texas and New Mexico into Arizona and the Sonoran Desert. Other moisture-laden air comes up from the Pacific Ocean to the south, originating either in the Gulf of Mexico or the southeastern Pacific. All these systems are the sources of summer monsoonal rains in the Sonoran Desert. The word "monsoon" is derived from the Arabic word for season and refers to a wind that changes direction. Winter rains are primarily from frontal storms moving east from the Pacific Ocean.

Chubascos

Chubasco is a Mexican term for the violent tropical storms, including hurricanes, that occur in the summer or fall and strike Baja California and the west coast of Mexico. Most form over the Pacific Ocean near the equator, but some originate on the east side of the continent and cross over to the Pacific. They can move at speeds of 12 to 18 miles per hour and have winds of 120 miles per hour or more. High waves caused by chubascos play a major role in dune-building or alteration along the coast. Winds, high waves, and heavy rains often cause serious flooding and property damage.

Although more common south of Kino Bay and at the southern tip of Baja California, an occasional chubasco reaches Kino Bay. Lester, a Class 1 hurricane in 1992, did extensive damage to homes at Kino Bay, especially those along the beach. Tropical storms that dump several inches of rain in a 24-hour period strike Kino Bay about every three to four years.

Temperatures

The Sonoran Desert is the hottest overall of the four Great Deserts. Higher elevations in the eastern and northeastern part of the desert, such as Tucson, are subject to winter frosts and brief periods of freezing temperatures, while portions along the Gulf and Pacific coasts, such as Kino Bay, are nearly frost free. The highest temperature recorded in the Sonoran Desert was 134 degrees at Sierra Blanca in the Pinacate Desert of northwestern Sonora in June 1971.

Why is it hot and humid at Kino Bay during the summer? The reason has to do with the temperature of the Sea of Cortés. Deserts next to warm oceans have summers that are humid. Examples are the east coast of the

Baja peninsula and the mainland bordering the Sea of Cortés. Deserts bordering oceans with cold currents, such as the west coast of the Baja peninsula facing the Pacific Ocean, have cooler and foggy summers. Summer temperatures there are about 15 degrees cooler than Kino Bay due to cooler breezes off the Pacific Ocean.

The Sky Islands

Most mountains in the Sonoran Desert are so low their slopes support only desert plants. Higher ranges such as the Santa Rita Mountains south of Tucson are called "sky islands." They support plants like oaks and conifers that are remnants of mesic plant communities that once dominated what are now deserts in southern Arizona and northern Mexico two million years ago. Sky islands adjoin, but are not a part of, the Sonoran Desert. As weather conditions became warmer and drier, the mesic plant communities disappeared from what is now the Sonoran Desert and moved up in elevation to cooler, moister habitats on the adjacent high mountains.

The sky island plant communities of today are an interesting link to the Sonoran Desert's past.

The Seven Regions of the Sonoran Desert

Forrest Shreve divided the Sonoran Desert into seven regions based on plant associations. He gave each a geographic (regional) name and a vegetative aspect name. A vegetative aspect is a visual impression of two or more perennial plants that characterize the region, but are not necessarily the most abundant species. For example, the plant aspect for the Central Gulf Region which includes Kino Bay is "Elephant Tree-Limberbush." The author has enlarged on Shreve's list of plants for several of the regions to include some more commonly seen.

Central Gulf Coast Region

This region is a strip of land on both sides of the Sea of Cortés including Kino Bay. Most of the area has a highly erratic rainfall pattern with an annual average of 5.5 inches or less, and often little or no rainfall for several seasons in a row. An exception is Guaymas with 7.5 inches. Mountains in Baja California tend to place Kino Bay in a rain shadow from storms originating in the Pacific Ocean.

A period of high temperatures and high humidity occurs from about mid-June to mid-October. Summer temperatures are ameliorated by onshore winds. The plant aspect is Elephant tree and Limberbush.

Typical mainland locations are Puerto Libertád, Kino Bay, and Guaymas. Typical Baja California locations are Santa Rosalia, Loreto, and La Paz.

Palo Blanco (*Lysiloma candida*). These beautiful white bark trees grow along the east side of Baja California from just north of Santa Rosalia, south to the tip of the peninsula. The only mainland population is a small grove north of San Carlos, Sonora. Photo near Mulege, Baja California, Sur.

Kino Bay Sub-Region

Kino Bay has an estimated annual average of 5.5 inches of rain; half falling in summer and half in winter. Rainfall is highly undependable. Despite this, there are more species of plants in the vicinity of Kino Bay than for most areas of similar size in the Sonoran Desert. This diversity is due to a large variety of habitat types such as dunes, estuaries, wetlands, mountains, bajadas and flats. In addition, some plants are favored by the maritime influence from the Sea of Cortés that brings cool breezes and higher humidity. Frosts are rare. Some

Red Elephant Tree (*Bursera hindsiana*)

northern gulf coasts of Baja California and Sonora. Elevations range from sea level to about 1,300 feet.

The Colorado Valley Region has the lowest precipitation and highest overall temperatures of the seven regions. Precipitation ranges from about 1.5 inches to 6 inches per year. Most precipitation falls in winter. It has fewer hills and mountains than the other regions, is the most arid, and has the lowest number of perennial plant species. However, there are a great many annuals that bloom following winter rains. The typical plant aspect is Creosotebush and White Bursage. Typical locations are Yuma and Phoenix, Arizona; San Felipe, Baja Norte; and Caborca and Puerto Peñasco, Sonora.

plants are able to utilize moisture from condensation of the coastal area's high summer humidity. Others are salt-adapted and able to utilize salty or alkaline subsoil moisture.

Common plants here are Saguaro and Cardón Cactus, Western Honey Mesquite, Ironwood, several species of palo verde, Ocotillo, dune succulents, several species of saltbush, and sarcocaulescent (fleshy trunk or limb) trees and shrubs. The latter include two species of elephant tree, two of limberbush, and the Boojum tree.

Lower Colorado Valley Region

This is the largest region. It is centered on the mouth of the Colorado River with portions in southwestern Arizona, southeastern California and along the

Seen from a distance, the shape and blue-green cast of Smoke Tree (*Psorothamnus spinosa*) give the impression of smoke rising far out in the desert. Smoke Tree is a shrub or small tree in the drier parts of the Lower Colorado Valley region. Photo taken north of Puerto Peñasco, Sonora.

Arizona Upland Region

This region is located in the northeastern corner of the Sonoran Desert in Arizona and Sonora. Elevations range from 500 to 3,000 feet. Precipitation averages range from 3 to 14 inches. Most rainfall occurs in summer. The higher rainfall and cooler conditions of the region have favored a large variety of plant species, high plant density, and vertical diversity (species with varying heights). It is called the "stem succulent desert" because of its abundance of columnar cacti. The plant aspect is palo verde, prickly pear, cholla, and Saguaro. Typical locations are Tucson, Lake Roosevelt, and Wickenburg, Arizona; and Sonoyta, Altar, and Magdalena, Sonora.

Plains of Sonora Region

This region centers on Hermosillo in Sonora, Mexico. It consists of a broad undulating plain with few hills. Average elevations range from 350 to 2,500 feet.

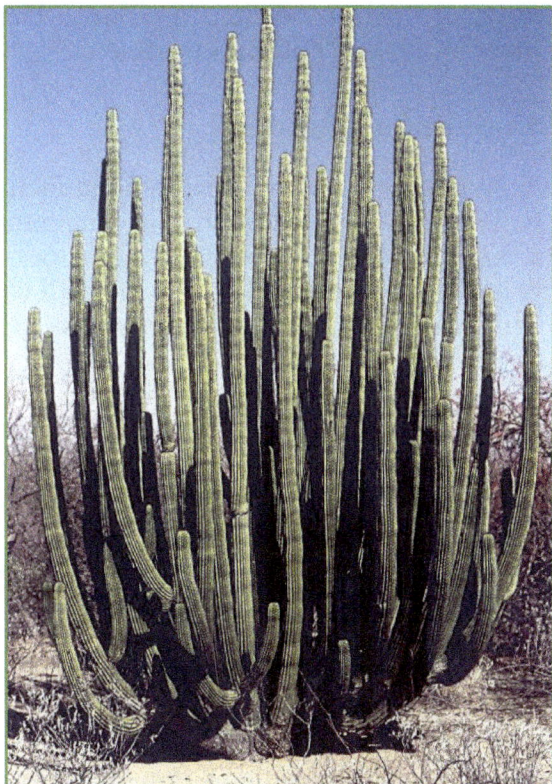

Organ Pipe Cactus (*Stenocereus thurberi*).

Precipitation ranges from 10 to 15 inches with most falling in summer. Frosts are infrequent. Only four mountains have elevations over 3,300 feet. Temperatures are intermediate between the Arizona Uplands and the Lower Colorado Valley Regions. Summer temperatures at Hermosillo often exceed 100 degrees.

Vegetation is dominated by a great variety of trees (includes columnar cacti) and shrubs. Organ Pipe and Senita cactus are common. A cross section of this region can be seen when driving south from Benjamin Hill to Hermosillo and west to Miguel Alemán (Calle Doce). Plant aspect is Ironwood-Brittlebush, with several species of columnar cacti and a large number of flowering shrubs and trees. Typical locations are Hermosillo, Miguel Alemán, Carbo, and Ures.

Foothills of Sonora Region

Located in Sonora, Mexico, this region joins the Plains of Sonora Region to the southeast. Elevations range from sea level to 3,500 feet. Rainfall exceeds 10 inches with a higher percentage of summer rain than the Plains of Sonora Region. The most notable features are the abundance of small trees, frequent groups of shrubs, heavy growth of herbs in summer, grass at higher elevations, occurrence of palms, infrequency of cacti, and the appearance of many trees and shrubs not found elsewhere in the Sonoran Desert. The area is the least desert-like of the regions. It is more of a transition zone into the thornscrub forests to the south, and some plant geographers have now eliminated it from their maps of the Sonoran Desert. The plant aspect is Acacia-Mesquite. Typical locations are Arizpe on the Rio Sonora, Obregón, and the lower Rio Yaqui.

Vizcaíno Region

The Vizcaíno is an area in central Baja California fronting the Pacific Ocean. The terrain is hilly or rolling with a few small mountains and many rough volcanic rock fields. The region is flanked to the north by the Sierra San Pedro Mártir, the highest mountain range on the peninsula.

Rainfall is predominately in winter followed by about eight months of very dry weather. Rainfall averages 5 inches or less. High winds from the Pacific Ocean buffet the coast, distorting the shape and reducing the density of plants there. The distinctive feature of the region is the presence of leaf-succulent plants including large species of *Agave* and smaller but more numerous species of *Dudleyas*. Tree Yucca, or Datillo, a giant *Yucca*, is common. Other notables are Boojum, Elephant Tree, and Cardón Cactus. Many species are favored by the maritime influence of the Pacific Ocean. It is here that Cardón and Boojum are most abundant and attain their greatest size.

Ball Moss (*Tillandsia recurvata*) is an epiphyte growing here on a Boojum tree. This "air plant" will attach to any tree species and is harmless to its host. Its presence is due to the maritime climate of the Pacific Ocean. It is common in the Vizcaíno Region.

The plant aspect is Agave and San Diego Bursage. Typical locations are Santa Maria, Guerrero Negro, and San Ignácio.

Magdalena Region

The region is a narrow strip of land facing the Pacific Ocean below the Vizcaíno region in Baja California. It is the southernmost region of the Sonoran Desert. Rainfall occurs predominantly in winter. Rainfall averages 4 to 5 inches.

The landscape is less striking than in the Vizcaíno Region due to the absence of Boojum and Elephant trees and the infrequent and poor development of Cardón Cactus, *Yuccas,* and *Agaves*. Prickly pears, Mesquite, Wolfberry, Sour Pitahaya, and Palo Blanco (*Lysiloma candida*) are common. Many plants growing here favor the Pacific maritime influence.

The plant aspect is Sour Pitahaya Cactus and Palo Blanco. Typical locations are San José de Comondú, Ciudad Constitución, and Agua Fresca.

Tree Yucca or Datillo (*Yucca valída*) is common in the Viscaíno Plain, often forming forests.

Some Ethnobotany of the Seri Indians

The Seri are native to the mainland portion of the Central Gulf Coast Region and Tiburón Island. From the book *People of the Desert and the Sea,* we know the Seri recognized and named 374 species of seed plants and used 94 of them for food. They were seafaring hunter-gatherers who once had an excellent diet from the desert and sea. Unlike most indigenous people in Mexico, they were not agricultural. The name Seri comes from a collective Spanish term meaning, "wild, and non-agricultural." The Seri call themselves, Comcaac or Kunkaak, meaning "people."

Historically, after wintering near the coast, the Seri broke into extended family groups and went inland for the native plant harvest. Mesquite beans were the primary food source. They ripened in late June or July and were picked off the tree when green or off the ground when dry. Mesquite was the single most important plant to the Seri. It was a primary source of food and also provided fuel, shelter, material for weapons, tools, fiber for ropes and nets, medicine, pestles for grinding food, carrying yokes, and many other things. The Seri Indians also used other native trees and shrubs for similar purposes.

Seri woman digging Coap (*Cnidoscolus palmeri*), an edible tuber. Spanish name is Mala Mujer (Bad Woman) because the stem and leaves have stinging hairs that are painful if touched.

Sugar-rich foods were eagerly sought. The main ones were the fruits of five species of columnar cacti (including Saguaro and Cardón), *Agave* hearts, Desert Wolfberry, and wild honey. Fruits of Saguaro and Cardón ripen in late June or July and were harvested with poles made of Saguaro ribs or the lateral roots of Mesquite. The pendant fruits of Chainfruit Cholla were abundant and could be gathered year-round.

Saguaro fruit is a tasty treat. The fruit contains numerous small black seeds embedded in a sweet juicy pulp that is brilliant red when ripe. The Seri removed the skin from the fruit, and the seeds and fruit were eaten fresh or the seeds could be dried and stored.

Later in the year, the Seri dug out pack rat nests for mesquite beans, cactus pods, and many other plant fruits and seeds.

Most Seri no longer harvest native plants for food but continue to harvest Saguaro fruit which they make into wine for a three-day celebration of the "new year" in early July.

Most ethnobotany information presented in this book, and the photo on this page, are from *People of the Desert and Sea: An Ethnobotany of the Seri Indians*, 1991, by Richard Stephen Felger and Mary Beck Moser ©1985, The Arizona Board of Regents. Reprinted by permission of the University of Arizona Press.

Invasive, Non-Native Plants

About 13% of the 2,500 species of plants in the Sonoran Desert are non-native. Many arrived by accident from the Old World and some were introduced as agricultural or garden plants. Others migrated here from various parts of the New World, carried by livestock, vehicles, improperly cleaned commercial seed, and a host of other avenues. The distribution of non-native plants in the Sonoran Desert has increased at a surprising rate. Researchers studying vegetation plots at the old Desert Laboratory on Tumamoc Hill just west of Tucson noted three non-native species in 1906. By 1983 the number had increased to fifty-six and is likely higher today. Buffelgrass, a non-native invasive recently placed on Arizona's noxious weed list, now infests 30 to 40% of the laboratory's 880 acres.

Many invasive plants are highly adaptive. They establish easily and spread rapidly.

Some have no natural control agents such as diseases or parasites in their new habitats, giving them an advantage over native plants. Many invasives get their foothold in disturbed sites, such as roadsides and abandoned fields, and spread from there.

Non-native invasives pose several threats to the ecosystems of the Sonoran Desert. Some out-compete and crowd out natives that are less competitive. Some, like Buffelgrass, grow dense and provide a ladder for fire to reach the canopy of fire-intolerant species like Saguaro, often killing entire stands. When invasives move in, the plant composition changes and so does the wildlife component. Despite efforts to curb these plants, most will continue to spread until they reach their range of adaptability and/or biological agents evolve to limit them.

About the Cactus Family

All cacti originated in the Western Hemisphere. The oldest cactus fossils were found in Utah and Colorado and date from fifty million years ago. Some cactus-like plants evolved in the Old World under similar conditions that favored evolution of cactus in the New World – a phenomenon known as convergent evolution.

The spines on cactus are actually modified leaves. Through evolution, the leaves of certain plants, including most cacti, were transformed into spines. A few species of cacti here at Kino Bay, such as Chainfruit

Cholla, produce some small, succulent, finger-like leaves early in the spring that drop off or become spines.

Soon after discovery of the New World, cactus specimens were sent to Europe to be classified and for planting. By 1737, Linnaeus, the famous Swedish botanist, combined the twenty-four known species of cacti into one genus he named "Cactus" from the Greek word "kaktos" meaning "the bristly plant." Many species of cacti were planted in the Old World where some became useful while others became nuisance weeds.

The Spaniards found Indians eating Nopal Cactus (*Opuntia ficus-indica*), a spineless prickly pear. Missionaries sent plants to Spain where Nopal was genetically improved into an even more desirable plant and later returned for planting in Mexico. Nopal is now planted all over Mexico, and the raw or canned pads are sold in grocery stores. Young pads of several species of prickly pear are eaten boiled or fried.

According to Aztec legend, an Indian Priest said they would wander until they saw a sign: an eagle with a snake in its beak sitting atop a prickly pear cactus. There they founded their capital, Tenochtitlán, at the present site of Mexico City. Following the revolution with Spain, Mexico adopted this symbol as the official seal and the emblem for the flag of the new republic – a fitting tribute to the indigenous people and the cactus family.

Transporting Plants out of Mexico

It is illegal to take live or dead plants or plant parts from Mexico into the United States. There are several reasons including: (a) to prevent the introduction of diseases or pests carried by plants, (b) to prevent the spread of non-native, invasive plants, and (c) to protect threatened, endangered or rare species. International regulations protect plants around the world that are threatened, endangered, or rare. (Washington Agreement to Protect Species, enacted 1973).

The Convention on International Trade in Endangered Species of Wild Fauna and Flora (CITES Act) specifically prevents the transfer of any cactus or cactus part (including cactus boots and cholla skeletons) from Mexico into the United States. Violations can result in fines.

If you have plants or plant parts in your vehicle when crossing the border into the United States, be sure to declare it to the U.S. Customs Agent. Better still – get rid of them and avoid a long delay from customs inspections of your vehicle.

The Plants

KEY TO ABBREVIATIONS:
spp. = Species
ssp. = Sub species
var. = Variety

Area A

Roadside Plants from Nogales to Kino Bay

A-1 TREE MORNING-GLORY, Palo Blanco, Palo de la Muerte, Palo Bobo (foolish tree)

Scientific name: *Ipomoea arborescens*
Family name: Convolvulaceae. Morning-glory family

A slender tree, to 40 feet tall, with an open canopy. It is the only tree in the morning-glory family. The bark is smooth and whitish. Stems exude a white latex when cut.

Tree morning-glory has large, white, flowers typical of the garden variety morning-glory. Flowers appear November through April when leaves are absent.

Leaves appear in spring after the flowers have dropped and persist into October. The large leathery leaves are ovate, medium green, and from 3 to 8 inches long.

Tree morning-glory is almost entirely confined to the State of Sonora, Mexico.

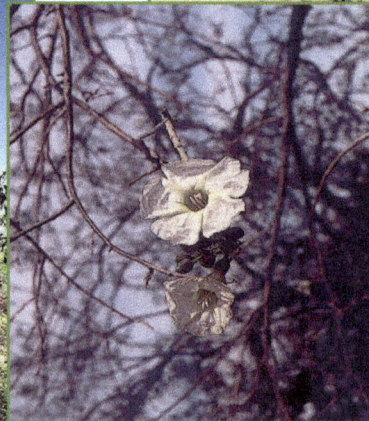

The flowers fall in late winter, and there is a long period in the spring when the tree is leafless, appearing dead, before the large, green leaves appear in late spring.

Tree morning-glory is an attractive tree that seems out of place in its desert environment.

WHERE TO SEE IT
Fairly common along Mexican Highway 15 just north of the Hermosillo toll booth. They also can be seen around Alamos and from the train approaching Copper Canyon from the west.

A-2 TREE OCOTILLO, Jaboncillo, Chunari, Torotillo
Scientific name: *Fouquieria macdougalii*
Family: Fouquieriaceae. Ocotillo family

A small tree to 25 feet tall with one to four short, greenish-yellow trunks and many smaller stems that resemble common Ocotillo. The light bronze colored bark of the trunk and stems is thin and exfoliates in sheets. Leaves grow following rains and are shed during dry periods.

The bright scarlet flowers appear at the end of stems any time from February to October following rains. Flowers are on long pedicels that cascade downward. See C-5, Ocotillo, for floral characteristics of three species of *Fouquieria* in our area.

Tree Ocotillo differs from common Ocotillo by having several distinct trunks and a tree-like growth habit. Common Ocotillo stems arise from the ground surface and are thin and wand-like. Although it has no trunk, common Ocotillo sometimes has the appearance of having a short trunk because erosion has washed soil away from the upper part of the root. Ranchers use the trunks of Tree Ocotillo for fence posts, and some sprout to form living posts.

The species was named for Daniel T. MacDougal, director of the former Carnegie Desert Laboratory at Tucson, who was an early botanist in the Sonoran Desert.

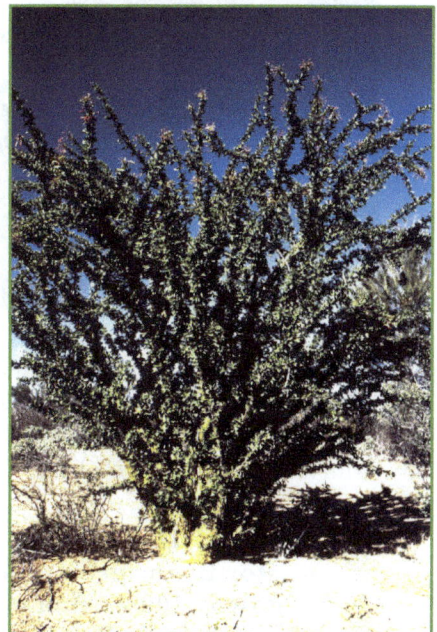

WHERE TO SEE IT
Tree Ocotillo is commonly seen along Mexico Highway 15 north of Hermosillo when in bloom beginning about February. The highway marks roughly the western extent of the species in Sonora, which shows its preference for that area of the Sonoran Desert with predominately summer rainfall.

A-3 CASTOR BEAN, Higuerilla

Scientific name: *Ricinus communis*
Family name: Euphorbiaceae. Spurge family

Castor Bean can be a forb, shrub, or tree. It can be an annual or a perennial. It is usually a shrub or a small tree to 10 feet tall, but may grow to 30 feet tall.

Castor Bean has very large leaves that are deeply cleft into seven to ten lobes with serrated edges. Leaves are usually dark green, but can be dull red, purplish, or bronze.

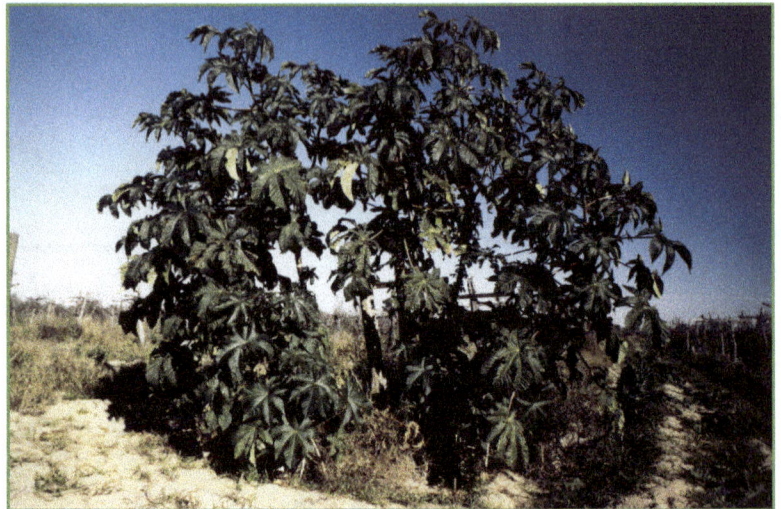

Castor Bean has male flowers beneath the female flowers. Female flowers have bright red styles. The pod is burr-like, green (photo lower right), turning bright red with age (photo left), and contains one seed resembling a tick. The scientific name "Ricinus" is also the name of a small brown tick that infests sheep and dogs. Clusters of the bright red pods at the top of stems are very attractive.

Castor Bean was introduced into Mexico from Africa where it was cultivated as a crop and landscape plant. It escaped cultivation and now grows wild where there is adequate moisture. Oil from the seeds is used as medicine, castor oil, lubricant for airplane engines, and to make plastic material. All parts of the plant are poisonous, especially the seeds. The poisonous property is the highly toxic phytotoxin, "ricin." Heating destroys the toxic property. Castor oil is not toxic despite what you might have believed as a child!

WHERE TO SEE IT
Castor Beans grow wild along Sonora State Highway 100 west of the town of Miguel Alemán (Calle Doce) toward Kino Bay.

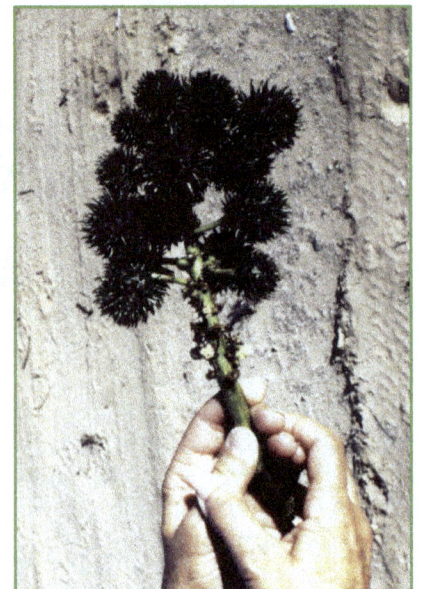

20

A-4 PRICKLY POPPY, Cardo, Nardo, Chicalote

Scientific name: *Argemone platyceras* var. *platyceras*
Family name: Papaveraceae. Poppy family

A tall (2 to 3 feet) perennial poppy with large white flowers growing in waste places especially along roadsides. Leaves and stems have prickles resembling thistles. Can be a weed on farm and ranchland. There are several species in Sonora that all look much alike. Widely distributed and common from Nebraska through the southwestern states, southern California, Baja peninsula, and into Mexico. Ranges in elevation from 1,500 to 8,000 feet.

The flowers are 2 to 3 inches across with white petals and yellow stamens in the center. The plants flower most of the year.

The name comes from the yellow prickles on the leaves and stems resembling a thistle; however, Prickly Poppy is a true poppy and not a thistle.

Another species, *Argemone ochroleuca*, is a smaller plant with pale yellow to orange petals. It occurs from the vicinity of Tucson, Arizona, south through Sonora and Central America into South America.

The Aztec Indians used the various species of *Argemone* for treating sore eyes. Mayo and Pima Bajo Indians of central and southern Sonora made a tea from the seeds for a purgative. The Seri Indians made a tea to cause expulsion of remaining placenta.

Tepehuan Indians used the milky juice from the stems to kill fleas and lice and as a topical application for burns.

WHERE TO SEE IT

Median strip and roadsides along Mexico Highway 15 between Nogales and Hermosillo. Its presence there is due to the favorable microclimate provided by rainfall runoff from the highway. Prickly Poppy is an invader in disturbed sites such as roadsides and abandoned or overgrazed fields where there is more moisture than the surrounding desert.

A-5 SINA CACTUS, Cina, Nacido, Tasajo

Scientific name: *Stenocereus alamosensis (Rathbunia alamosensis)*
Family name: Cactaceae. Cactus family

A columnar cactus, 6 to 8 feet tall, with stems 2 to 4 inches in diameter, forming large, nearly impenetrable colonies as much as 40 feet across. Plants in the center of the group gradually die out leaving an opening in the old colony. The stems are often sprawling or the tops are tipped over. Plants reproduce when tops bend down, touch the ground and take root. It appears that reproduction from seed is rare.

Fruits are round, bright red, and about 2 inches in diameter. The ripened fruit has split open.

The Seri Indians considered the fruit to be bitter. It was sometimes eaten fresh but not harvested for storage.

Sina Cactus is a native of Mexico. It shows a decided preference for areas of predominantly summer rainfall. Its western-most limit in our area is several colonies visible from Sonora State Highway 100 between kilometer markers 45 and 50 between Hermosillo and Miguel Alemán (Calle Doce). It does not grow at Kino Bay.

Sina Cactus is closely related to and resembles Pitahaya Agria *(Stenocereus gummosus)* shown in part C of this book. A few colonies of Pitahaya Agria can be seen near Estero Santa Rosa just north of Kino Bay.

WHERE TO SEE IT

A colony can be seen between Miguel Alemán and Hermosillo on the south side of Highway 100 opposite kilometer marker 50 (see photo above left). It also can be seen along Mexico Highway 15 north of Hermosillo between kilometer markers 30 and 80.

Scarlet, tubular flowers, 3 to 4 inches long appear from March through May and again from July through August. Flowers remain open all day. Hummingbirds, on their northward spring migration, are the chief pollinators.

A-6 BUFFELGRASS, African Foxtail Grass, Zacate Bufel, Bufel de India

Scientific name: *Pennisetum ciliare*
Family name: Poaceae. Grass family

A light green, perennial, dense-growing, bunchgrass averaging 20 to 30 inches tall on dry land conditions and to 6 feet tall with irrigation. Root systems are usually stoloniferous. Leaf blades are bluish-green with soft hairs on the surface. Leaves turn brown during dry periods and are replaced with green leaves after rains. This grass is easily identified by its light-purple foxtail head. When back-lit by sunlight, the light purple cast of the seed head is very attractive.

Buffelgrass is drought tolerant and regrows following a little rainfall any time of year. Seed heads produce large numbers of wind-dispersed seeds which establish easily almost anywhere. The plant requires summer moisture and is not cold tolerant. It is not to be confused with "Buffalo grass," a native of the American Midwest.

Buffelgrass is a native of Africa and a cousin to Fountain Grass *(Pennisetum setaceum),* a common ornamental in the American Southwest. Buffelgrass was brought to the United States by the USDA Soil Conservation Service and developed as a pasture and dry land range grass. It is a popular grass in Texas where it was used to revegetate thousands of acres of rangeland. It was introduced into Mexico about 40 years ago where it has the potential to triple forage production over native forage plants. Over a million acres of desert and sub-tropical thornscrub land in northern Mexico have been cleared and planted to this grass. The benefit or harm depends on one's point of view. It has been a boon to the cattle industry, but a dangerous threat to native plants. The State of Arizona listed Buffelgrass as a noxious weed in 2005.

Many species of plants in the Sonoran Desert developed over thousands of years in the near-absence of fire because plants or plant groups are widely-spaced, leaving natural barriers to fire. Because of its competitive nature, Buffelgrass develops dense colonies that gradually crowd out less competitive native plants. Buffelgrass provides a ladder for fire to reach the canopies of fire-intolerant plants such as saguaro, often killing entire stands. Some scientists believe fire effects from Buffelgrass could change cactus forests into grasslands within several years of introduction.

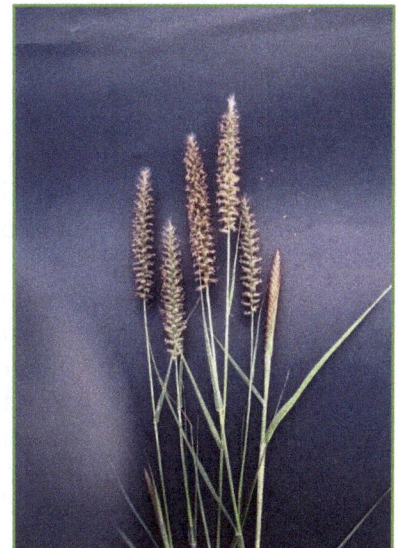

WHERE TO SEE IT
Buffelgrass is common along
Mexican highway 15 north of Hermosillo and Sonora
State Highway 100 between Hermosillo and Kino Bay.

A-7 MEXICAN PALO VERDE, Horsebean, Jerusalem Thorn, Bagote, Guacoporo, Retama

Scientific name: *Parkinsonia* (*Cercidium*) *aculeata*
Family name: Fabaceae (Leguminosae). Pea family
Subfamily: Caesalpinioideae. Senna subfamily

A graceful tree, usually 20 feet tall (to 40 feet) with rounded crown; short trunk; thin, light green limbs; and long, drooping pinna of leaflets. Bark of the trunk turns dark brown with age.

The most outstanding feature of the tree are the long (12 to 14 inch), graceful pinna (singular) which is a leaf composed of a rachis with ten to forty pairs of tiny leaflets. The leaflets fall with the first sign of water stress, leaving the rachis to serve as the chief photosynthetic organ. The leaflets are 2 to 8 mm long and seldom survive more than five to ten weeks. The rachis is shed during severe drought. Mexican Palo Verde can easily be distinguished from other species of palo verde which have much shorter pinna.

Stems have one prominent spine (1 to 3 cm long), and two opposite, shorter spines at the base of each cluster of pinnae (plural). The short spines may eventually drop off. The long spine was a leaf stalk that has become a spine. This occurs only on the current year's growth of a stem. Pinnae that arise from the older part of the stem develop just above the old spines.

Mexican Palo Verde is a native to the New World tropics that was later introduced as a cultivar worldwide and has become naturalized over much of the Sonoran Desert. It is closely related to Little-Leaf Palo Verde, Blue Palo Verde, and Palo Brea, or Tar Tree, and often crosses with them to produce confusing hybrids. All four species occur in the area covered by this book.

The tree flowers March through August in the northern (Arizona) part of the Sonoran Desert, and April through May in the southern (Kino Bay) part. Numerous small, half-inch, bright yellow flowers cover the tree. Four of the five petals are yellow. The fifth, or banner petal, is yellow and develops red spots or turns orange with age. (Note: The banner petal in most species of the pea family is distinctly larger than the other four petals.)

The Palo Verde Beetle, also known as the Palo Verde Root Borer (*Derobrachus hovorei*), is a large, beetle that feeds on Mexican Palo Verde and other non-native trees, but not other

24

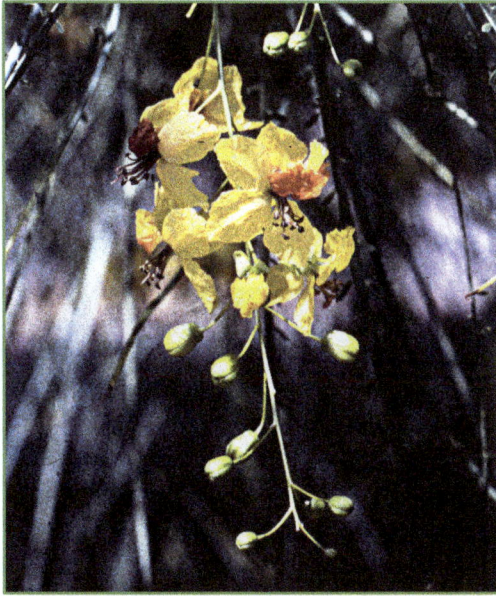

species of *Parkinsonia*. It spends several years as a large underground grub feeding on the roots of the tree. When summer rains arrive, adults emerge as beetles and fly in the early evening hours in search of mates. They are attracted to outdoor lights. Adults are black, 5 to 6 inches long, have very long antennae, big eyes, and large mandibles.

Mexican Palo Verde is not easily killed by fire and readily resprouts following top-kill. It can regrow as much as 8 feet in one year. It is a prolific seeder, and seedlings establish and grow quickly. Young branches are often cut to feed livestock.

Mexican Palo Verde is a popular landscape tree. It grows fast and makes a beautiful shade tree, but requires a large area when mature.

HOW TO TELL THEM APART:

There are five species of *Parkinsonia* in our area of Sonora; three are covered in this book. Blue Palo Verde and Palo Estribo are not covered but are discussed below.

1. Mexican Palo Verde *(P. aculeata)*: Leaf pinna is 12 to 14 inches long. Other species have pinna less than 3 inches. It has three straight spines at nodes on the stems.

2. Little-Leaf Palo Verde *(P. microphylla)*: Twigs are spine-tipped only. No spines at stem nodes. Leaflets tiny, 1 to 2 mm long.

3. Blue Palo Verde *(P. florida)*: Leaflets 5 to 9 mm long, usually three to four pairs per pinna. Thorns short, 4 to 10 mm long. One spine per node. A large tree usually growing along desert washes.

4. Tar Tree *(P. praecox)*: One or two spines at each node along stems. Spines stout, 5 to 20 mm long. Pinna has four to six pairs of leaflets. Leaflets 4 to 13 mm long. Pods flattened, not constricted between seeds.

5. Palo Estribo *(P. x sonorae)*: This tree is a hybrid between Little-Leaf Palo Verde and Tar Tree. It occurs where the ranges of the two overlap, including the area between Miguel Alemán and Hermosillo. Pods slightly constricted between seeds. Spines none or one and slender (not stout), 5 to 15 mm long. Leaflets 2 to 6 mm long. Pods constricted between seeds. Compare the characteristics with Tar Tree.

WHERE TO SEE IT

Mexican Palo Verde is very common along the highway east and west of Miguel Alemán (Calle Doce). In our area, it grows almost entirely along roadsides and sites with higher moisture than the surrounding desert. It is a common ornamental in Kino Bay. Look for the long (12 to 14 inch) pinna that distinguishes it from other palo verdes.

25

A-8 LITTLE-LEAF PALO VERDE, Foothills Palo Verde, Yellow Palo Verde, Paloverde, Dipua

Scientific name: *Parkinsonia* (*Cercidium*) *microphylla* (*microphyllum*)
Family name: Fabaceae (Leguminosae). Pea family
Subfamily: Caesalpinioideae. Senna subfamily

Little-Leaf Palo Verde grows only in the Sonoran Desert. It is a small tree to 25 feet tall with either a rounded or sprawling growth form. Four of the five flower petals are pale yellow; and the fifth, or banner petal, is creamy white. The cylindrical pods are 3 inches long and ¼ inch wide with long narrow points at each end.

The tiny leaflets are yellowish green, 1 to 2 mm long, arranged along the pinna in five to seven pairs. New leaves are produced in response to summer and winter rain. Winter leaves are shed before the trees begin to flower in April.

The tree has smooth green bark throughout except for the base of the trunk which is roughened and gray. During the dry season when the leaflets have dropped, photosynthesis is carried on by chlorophyll in the bark. The tree is spineless except for its spine-tipped branchlets.

The tree is widely distributed over much of southern Arizona, Sonora (north of Guaymas), and the Baja peninsula. It is replaced in riparian corridors by Blue Palo Verde.

Both Little-Leaf and Blue Palo Verde are Arizona's official state trees. Little-Leaf Palo Verde may hybridize with Blue Palo Verde in the northern part of its range (Arizona) and with Tar Tree in the southern part (Sonora). These hybrids produce combinations of characteristics that can be confusing to the "hip pocket botanist." The hybrids are popular cultivars found in nurseries and landscapes in southern Arizona.

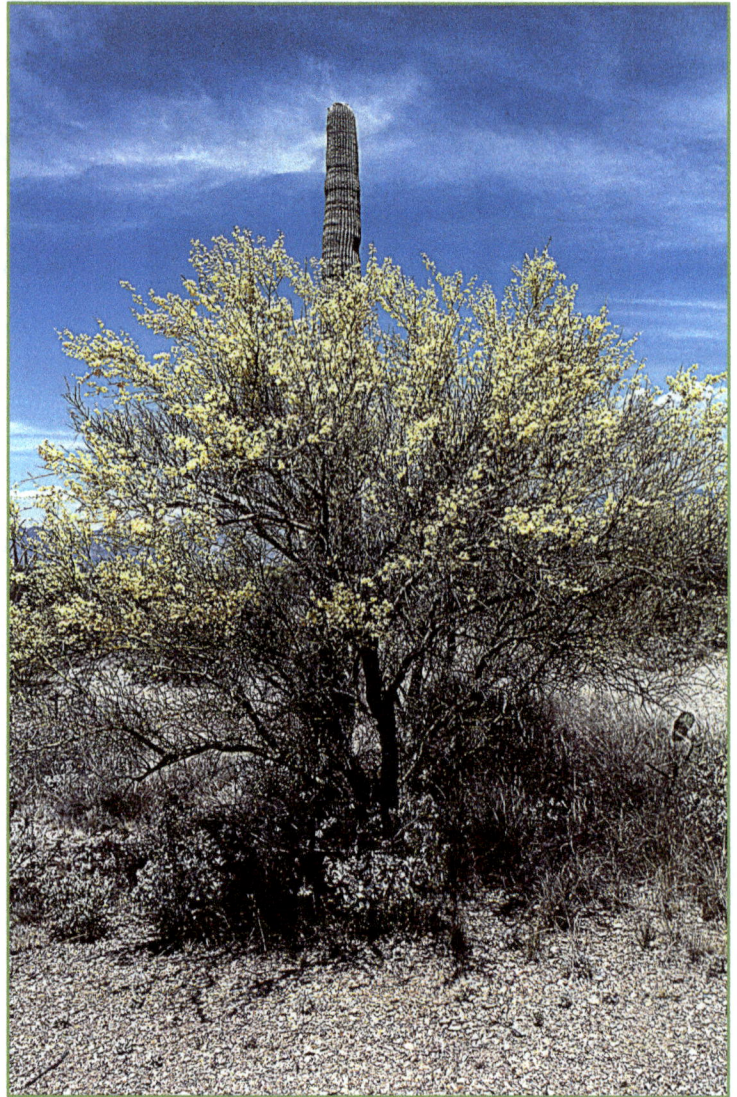

WHERE TO SEE IT
Little-Leaf Palo Verde is a common volunteer tree in vacant lots in New Kino and along Mexico Highway 15 north of Hermosillo. The key feature is that Little-Leaf Palo Verde is spineless. See HOW TO TELL THEM APART in A-7, Mexican Palo Verde, page 25.

A-9 GUAYACÁN, Arbol Santo, (Saint's tree)

Scientific name: *Guaiacum (Guajacum) coulteri*
Family name: Zygophyllaceae. Caltrop family

When in bloom, the dark blue flowers of Guayacán are striking. It is the most beautiful flowering tree of the Sonoran Desert. A small tree (15 to 20 feet) resembling Ironwood. It has dense, crooked branches, small dark green leaflets and very hard wood. Stems are thornless. Most trees have an irregular-shaped crown with no particular form. They grow very slowly and are not attractive or even recognized among Mesquite and Ironwood until in flower. Guayacán is subject to special protection in Mexico.

Guayacán distribution is limited. It is found from Sonora to Puebla and Oaxaca, and the interior of Tiburón Island. Mexico Highway 15 marks roughly the western extent of the species in Sonora. The trees on Tiburón Island may have been planted by Seri Indians who made their famous "Seri blue" paint from its resin.

The brilliant, indigo blue flowers are about an inch wide. Flowering may occur at any time from April through November, depending on rains; but the peak bloom is usually May to July. Flower color fades over a short time to light purple.

The fruit is winged and looks like a small purplish nut about the size of a marble. The leaves are pinnate with six to ten crowded leaflets ½ to 1 inch long. Leaves are drought-deciduous, often dropping in October. New leaves appear in spring.

Seri Indian women made a blue dye from the tree resin mixed with white clay and the boiled ingredient of the root bark of White Bursage *(Ambrosia dumosa)*. The formula was a secret known only to the women. Girls were taught the process when they reached a certain age. The dye was used as face paint and to decorate pottery, arrows and many other things. For some reason, it was not used as a dye for baskets.

In the Seri supernatural, "Seri blue" paint and wood powder from Sonoran Caper *(Atamisquea emarginata)* were used for seeking good luck.

WHERE TO SEE IT

There is a large Guayacán north of Hermosillo in a turnout near kilometer marker 100 of the south-bound lane of Highway 15. Another tree is on the same highway in a group of white granite boulders on the east side of the north-bound lane near kilometer marker 25. Guayacán is easily seen when in flower.

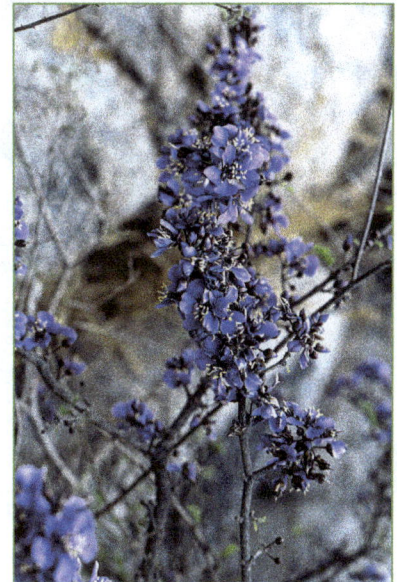

A-10 SUMMER POPPY, Arizona Caltrop, Arizona Poppy
Scientific name: *Kallstroemia grandiflora*
Family name: Zygophyllaceae. Caltrop family

An annual forb, 8 to 12 inches tall, with sprawling to upright, hairy stems and bright orange, poppy-like flowers. Following sufficient rain, it can be seen growing in dense patches along highways in Arizona and Sonora in mid-summer. The plants with their colorful flowers are most abundant in desert grasslands. It is not a true poppy. Summer poppy is closely related to Puncture Vine or Goathead (*Tribulus terrestris*), an introduced weed from Europe that is common around Kino Bay. Unlike the latter, Summer Poppy does not have burs.

Summer Poppy has large orange flowers with five petals and purple centers. Some petals may be pale orange with a deep orange spot at the base. Each flower is open only part of the day. Flowering occurs February to September, depending on the area. Here in the Sonoran Desert it flowers only in the summer – usually in August around Hermosillo. In early morning, new flowers are bright orange, fading to pale orange in the afternoon heat. In the southern part of its range, pink flowers may appear. Leaves are pinnate, with five to nine pairs of oblong leaflets. Leaves and stems look very much like Goathead weed.

Summer Poppy may be confused with California Poppy, also called Mexican Gold Poppy (*Eschscholzia californica* subspecies *mexicana]*, a true annual poppy that also grows in the Sonoran Desert. California Poppy usually has four petals and blooms in the spring. Summer Poppy has five petals and blooms in the summer here in the Sonoran Desert.

Summer Poppy grows in response to summer rains. Seeds are coated with a germination inhibitor that must be completely washed off before they will germinate, a process that requires rains for several seasons. This feature allows many seeds to stay dormant for years, extending species survival. The plants are present every summer, and abundant following winters with heavy rainfall.

WHERE TO SEE IT
Look for Summer Poppies in July through August along Sonora Highway 100 just west of Hermosillo and Mexico Highway 15 north of Hermosillo. It does not grow at Kino Bay.

A-11 VELVET-POD MIMOSA, Mauto, Gatuño

Scientific name: *Mimosa dysocarpa*
Family name: Fabaceae (Leguminosae). Pea family
Subfamily: Mimosoideae. Mimosa subfamily

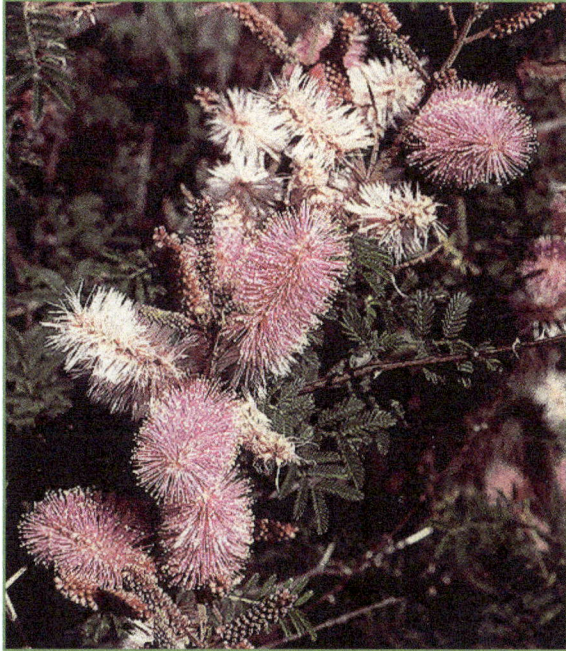

Velvet-pod Mimosa is a small, upright or sprawling shrub, 2 to 4 feet tall, with dense branches and short (¼ inch) straight or slightly curved thorns. Prefers well-drained, rocky soils in zones with about 10 inches of annual precipitation. Does not grow near Kino Bay.

Velvet-pod Mimosa has beautiful flowers with rose-pink stamens in 2 to 3 inch long cylindrical spikes. Flowering occurs July through September and occasionally as early as May or as late as October, depending on rainfall abundance. As the pink color fades to white, the shrubs become bi-colored. Stamens give the flowers their color and bottlebrush look.

The shrub has tiny leaflets typical of many plants in the *Mimosa* genus.

Velvet-pod Mimosa is a landscape shrub in southern Arizona.

WHERE TO SEE IT
Rocky road shoulders along Mexican Highway 15 between Benjamin Hill and Imuris between kilometer markers 224 and 226 and roadsides along Mexican Highway 2 east of Imuris.

A-12 FAIRYDUSTER, Huajillo, Mesquitilla

Scientific name: *Calliandra eriophylla*
Family name: Fabaceae (Leguminosae). Pea family
Subfamily: Mimosoideae. Mimosa subfamily

Fairyduster is a scraggly, thornless shrub, 2 to 4 feet tall, with no definite form. It grows on slopes and mesas and along washes in drier parts of its range. It establishes quickly on disturbed sites such as roadsides with well-drained, sandy or gravelly soil.

Flowers are a beautiful pink, rose, or purple powder puff that eventually fade to white, giving the impression of two colors of flowers. The 2-inch wide puff is actually dozens of slender stamens. Fairyduster may flower any time of year. Typical spring flowering starts in February at lower elevations and in March or April higher up. Plants occasionally bloom again in September and October. Leaves consist of many tiny (2 to 3 mm long), dark green leaflets, which wilt during dry periods and regrow following rains. It is used occasionally as a landscape plant.

Livestock and many desert animals feed on the foliage, and insects and hummingbirds frequent the flowers.

Baja Fairyduster *(C. californica)*, a close relative, is a larger shrub, with larger, dark red flowers. It was previously known only from the Baja peninsula and several gulf islands, including Tiburón, but has recently been found growing on the mainland near the coast north of Kino Bay. Baja Fairyduster, with its strikingly-beautiful, large red flowers, is a common landscape plant in Tucson.

WHERE TO SEE IT
Fairyduster is common along much of Mexico Highway 15 north of Hermosillo, growing on sandy or gravelly soils. It can also be found along the Minas Pilares mine road which branches west from Highway 15 just south of the Hermosillo toll booth. Also grows at Western Horizons R.V. Court 9 miles north of Kino Bay.

A-13 SANTA RITA PRICKLY PEAR, Duraznilla

Scientific name: *Opuntia santa-rita (O. violaceae var. santa-rita)*
Family name: Cactaceae. Cactus family

A prickly pear with short, stout trunk, growing to as much as 5 feet tall. The pads are round to broadly ovate, about 8 inches across, purple or purplish, with few or no spines. Spines, when present, will usually be along the outer margin of the pad, and possibly a few on the pad's flat surface. Spine length is variable, up to 3 inches long. Some pads are greenish-blue with purple spots around the aeroles.

Santa Rita Prickly Pear is common in south central Arizona (Tucson) and has a scattered distribution on each side of Mexico Highway 15, chiefly north of Benjamin Hill. A few have been found growing on rocky slopes of the mountains near Tastiota.

Flowers are pale yellow; to 3½ inches wide, followed by a fruit that is red to purplish, smooth, and slender. The fruit is spineless, although they

usually contain glochids. Flowering occurs April and May. The fruits (tunas) are processed to make wine, jam, and jelly. The Seri Indians made a pink or pale red face paint from the fruit.

Why don't we have prickly pear cactus (including Santa Rita) at Kino Bay? Prickly pear species are more sensitive to prolonged drought than any other type of cactus and are absent from extensive stretches of the Sonoran Desert.

WHERE TO SEE IT

The pictures were taken in the cactus garden at the toll booth in Magdalena, Sonora. Some Santa Ritas can be seen on the west side of Mexico Highway 15, between kilometer markers 94 and 97 of the south-bound lane. Santa Rita Prickly Pear is a popular landscape cactus in Arizona and Mexico, including Kino Bay. Look for the purple cast and sparse spines clustered near the upper edge of the pad. None of the prickly pear species are native to the Kino Bay area.

A-14 COCHINEAL SCALE, Cochineal Louse

Scientific name: *Dactylopius spp.*
Family name: Dactylopiidae

Several insects have evolved with cacti as their sole food source. Cochineal is a scale insect that feeds on prickly pears. The female attaches to the pad and forms a cottony mass around its body for protection. Females are purplish-gray, wingless, and legless. Mature males are winged and look like small white gnats. Females lay eggs that hatch into tiny nymphs that crawl or are blown to other pads to settle.

The body fluids of the female contain carminic acid, a bright crimson substance that was used by native people in Mexico and South America as a fast dye for coloring fabrics. Pima Indians in the American Southwest used it to dye the ends of war arrows.

When the Spanish found the Indians using this marvelous color, some dye was shipped to Spain where it became the royal crimson, reserved by law for the King's use only. Because it was thought that prickly pears produced cochineal, the plants were shipped to Spain, where they were planted and tended but failed to produce cochineal. Eventually, it was realized that cochineal was a parasite and must infect the host.

Later, production and export of cochineal dye became a major economic activity in Spain and Mexico. Its source was a secret for many years. Eventually, competitors discovered the secret and established populations of prickly pears in countries with dry climates, including India, Ceylon, South Africa, and Australia. Production proved profitable; but in some cases, the cochineal died and the prickly pears spread to become serious pests.

The cochineal industry thrived until the late 1800s when cheaper dyes became available. It is still manufactured in several countries, including Mexico. Modern handcrafters who practice dying wool with natural materials use this dye. They collect and dry the female insects, then immerse them in boiling water to release the dye. Cochineal is one of a very few red dyes approved by the U.S. Food and Drug Administration. It is currently used in red candies, beverages, and lipstick.

Sources: *A Natural History of the Sonoran Desert*, Arizona-Sonora Desert Museum, Tucson and *Desert Plants*, Vol. 6, No. 1, University of Arizona Press. Tucson.

WHERE TO SEE IT
Cochineal grows on several species of prickly pear in Sonora, Arizona, and New Mexico. It is common in some areas and totally absent in others. Look for the cottony balls. Scoop one off with a knife and smash it to see the brilliant crimson color.

A-15 Orchid Vine, Gallinita

Scientific name: *Callaeum (Mascagnia) macropterum*
Family name: Malpighiaceae. Malpighia family

An erect shrub, 3 to 7 feet tall, but more often, in our area, a vine which climbs on a host bush or tree. Vines and pods often nearly cover their host. Leaves are ovate to oblong, 2 to 22 mm wide and 1 to 3 inches long.

Numerous small, yellow flowers are followed by inflated four-winged, light green pods. The pods are 2 inches wide and slightly more than 2 inches long. At this stage, they can be gathered to make an attractive dry flower arrangement. The green pods turn brown with age and remain on the vine for several months. The pods are called "samaras," a technical term meaning a dry, indehiscent, usually one-seeded, winged fruit. Orchid Vine flowers any time of year following rain.

In Spanish, gallo means "rooster," gallina means "hen," and gallinita means "little hen." Apparently the Spanish name refers to the shape of the pod which resembles the inflated comb of some breeds of exotic chickens.

Orchid Vine is an occasional plant often found in disturbed sites such as roadsides and washes.

Orchid Vines and pods nearly cover a 10 foot tall host tree along Highway 15.

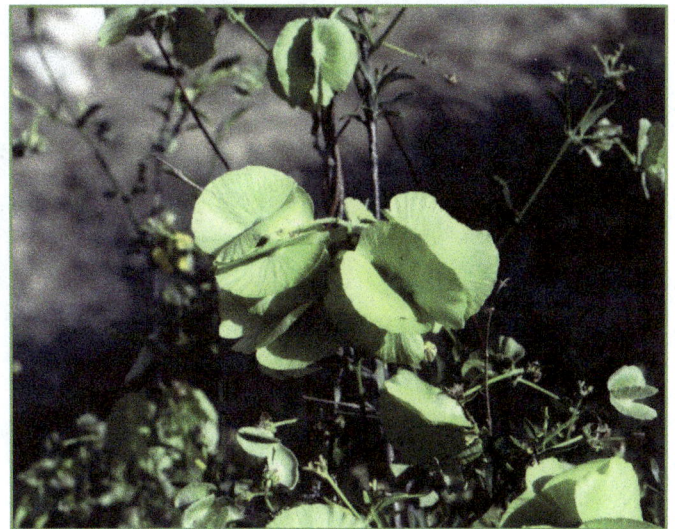

Orchid Vine pods

WHERE TO SEE IT
Rather common in places along State Highway 100 just west of Hermosillo. Look for it growing on fences in the vicinity of the airport. Also fairly common along Mexico Highway 15 north of Hermosillo. Although Kino Bay is too dry for Orchid Vine, an occasional plant can be found growing in arroyos near town.

A-16 KIDNEYWOOD, Palo Dulce

Scientific name: *Eysenhardtia polystachya* var. *tenuifolia*
Family name: Fabaceae (Leguminosae). Pea family
Subfamily: Papilionoideae. Papilionoid subfamily

Kidneywood is a tall shrub or small tree, usually 6 to 8 feet tall in our area. Leaves are pinnate, 2 to 6 inches long with twenty-one to forty-one oblong leaflets, ¼ to ¾ inches long. The undersurface of leaflets is conspicuously dotted with brown glands. Flowers are in dense spike-like racemes, 2 to 5 inches long, with numerous tiny white flowers about ¼ inch long. The flowers have a sweet scent. Stems are slender and thornless.

Flowering occurs sporadically from April to September. During years with heavy summer rainfall, the shrubs have an abundance of beautiful white flowers that appear like large catkins at the tops of plants.

Kidneywood's westward distribution in the Sonoran Desert ends at approximately the location of Mexico Highway 15, showing its preference for the areas where most precipitation comes in summer.

Distribution includes Sonora (mostly east of Highway 15), southwest Chihuahua, Sinaloa, southwestern New Mexico, and southeastern Arizona as far north as Tucson. In habitats such as oak woodland and the tropical deciduous forest around Alamos, Kidneywood reaches tree size of 20 feet.

Kidneywood is easily grown from seed or cuttings and makes rapid growth with irrigation. It makes a nice landscape tree.

WHERE TO SEE IT
Kidneywood can be seen growing along Mexico Highway 15 north of Hermosillo in habitats with a large number and variety of mature mid-to-tall shrubs and trees. It also establishes on disturbed sites such as the right-of-way along Highway 15 just south of Benjamin Hill.

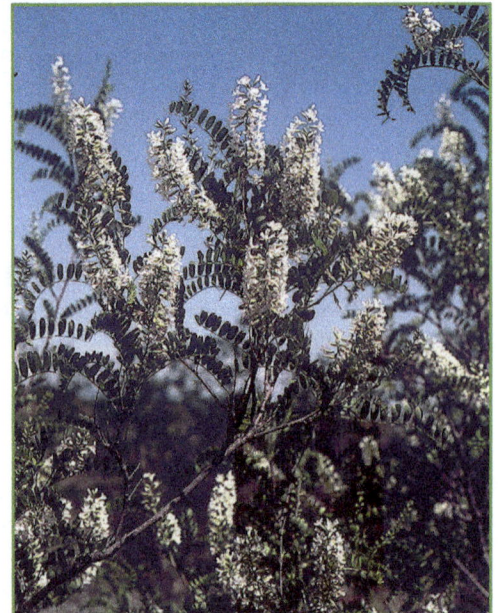

34

A-17 WHITE-BALL ACACIA, Prairie Acacia, Fern Acacia, Cantemo, Guajillo, Palo de Pulque, Barbas de Chivo.

Scientific name: *Acacia angustissima*
Family name: Fabaceae (Leguminosae). Pea family
Subfamily: Mimosoideae. Mimosa subfamily

A thornless shrub or small tree to 13 feet tall, but mostly 4 to 7 feet tall in our area. During flowering season, the shrub is covered with little white snowball-like flowers. It is a highly variable species with numerous varieties.

Flowers are white, sometimes tinged with pink, with numerous stamens crowded into a ½ inch ball. Flowers are in clusters along the stems, in leaf axils, and in elongated terminal clusters. Pods are brown, flattened, and up to 3 inches long. Flowering occurs April through October.

Leaves are green to bluish-green with many tiny leaflets. The stems are deeply grooved and may be glabrous or very hairy.

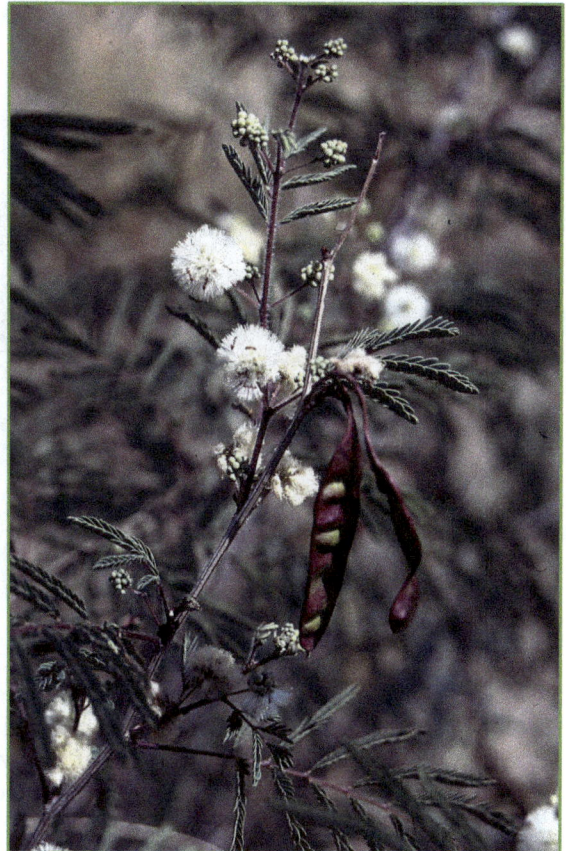

White-ball Acacia ranges from Missouri and Kansas southward into Central America at elevations from sea level to 6,500 feet. Populations are scattered in Sonora but are much more common in southeastern Arizona where it ranges from 3,000 to 6,500 feet in elevation. The distribution shows a preference for summer rainfall. It does not grow in the vicinity of Kino Bay and is absent from California and the Baja peninsula.

WHERE TO SEE IT
Common along Mexico Highway 15 in the vicinity of Benjamin Hill.

A-18 DESERT COTTON, Wild Cotton, Algodoncillo (little cotton)

Scientific name: *Gossypium thurberi*
Family name: Malvaceae. Mallow family

A tall, sparsely-branched, spindly tree to 14 feet tall, but usually much shorter in the area along Mexico Highway 15 south of Nogales. Trunk and stems are narrow.

Flowers are cup-shaped. Petals are about an inch long, white or cream-colored, with a faint purple tinge at the base, turning purplish with age. Petals may be black-dotted. The fruit is broadly ovoid to subglobose, 1 to 2 cm long, gland-dotted, and bearing tufts of long white hairs (cotton). The seeds are 4 to 5 mm long, dark brown to nearly black. Flowering occurs August through November. The flowers are quite pretty.

The leaves are deeply three- to five-lobed. The lobes are 1 to 3 cm wide and 3 to 10 cm long, lanceolate, gland-dotted, bright green above, and usually slightly glaucous beneath.

Desert Cotton grows along arroyos, canyons, and rocky hillsides. It is known from the Superstition Mountains near Phoenix and the Rincon Mountains near Tucson of southern Arizona, south to the states of Sonora, Chihuahua, and Jalisco in northern Mexico.

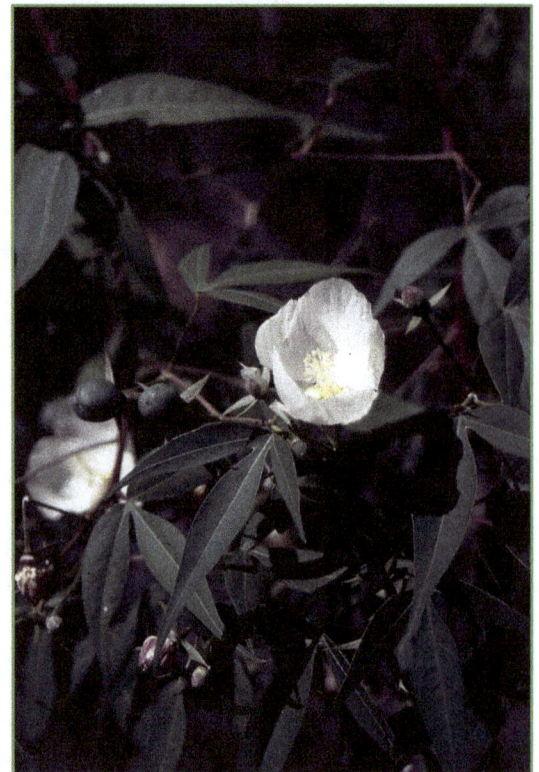

WHERE TO SEE IT
Desert Cotton is an occasional small tree growing on cut and fill slopes along Mexico Highway 15 north of Imuris. Look for it between kilometer markers 218 and 220 at the foot of bluffs on the east side of the highway. The large white flowers are easily seen from the highway.

A-19 COURSETIA, Baby Bonnets, Chino, Samo, Samo Prieto, Causamo, Mountain Samo, Samota, Cousamo, Chipile, Chipillo, Tepechipile, Guarijio

Scientific name: *Coursetia glandulosa (C microphylla)*
Family name: Fabaceae (Luguminosae). Pea family
Subfamily: Papilionoideae. Papilionoid subfamily

A thornless shrub or small tree to 20 feet tall, but more commonly a shrub 5 to 8 feet tall in our area. The many flexible, arrow-like stems arising from the base of the plant are distinctive and make it easy to spot. The bark is smooth, gray, and faintly striate.

The flowers are pea-like, to ½ inch long, single or in groups with cream-colored upper petals and yellow lower petals. The banner petal often has a faint tinge of red. Flowering occurs January through February in southern Sonora; February through March in northern Sonora; and March through April in Arizona. Coursetia will usually have an abundance of flowers following rain. Flowers and leaves appear about the same time.

The fruit is a brown, thick-walled pod, to 3 inches long, that is flattened between the seeds.

The pinnate leaves are 1 to 2 inches long with four to nine pairs of leaflets. Leaflets are 0.5 to 4 cm long, thin, oblong to oval, and often with a point at the tip.

Coursetia is widely distributed in Sonora. It also occurs in south-central Arizona, the cape region of Baja, and from Chihuahua to Oaxaca.

The wood is flexible and very hard. Seri Indians favored the stems for making bows. It was also used for making the loops for cradle boards and for digging sticks. An orange lac is deposited on the stems by *Tachardiella*, a scale insect. It was used as a glue and sealer by native people. Sonoran pharmacies once sold it as "goma Sonora" for use in treating colds, fevers and tuberculosis.

WHERE TO SEE IT
Coursetia is rather common in rocky road cuts along Mexico Highway 15 south of Benjamin Hill. Look for it near the pull-out at kilometer marker 100 in the south-bound lane. Does not grow at Kino Bay.

A-20 BIRD'S FOOT MORNING GLORY

Scientific name: *Ipomoea ternifolia* var. *leptosoma*
Family name: Convolvulaceae. Morning glory family.

An annual, twining or spreading vine with attractive, trumpet-shaped flowers. It grows on hillsides, arroyos, and grassy plains in southern Arizona, New Mexico, Sonora, and south to Oaxaca. It also grows on the Baja peninsula from vicinity of San José de Comondú and Bahía de la Concepcíon to the Cape Region. It is an occasional plant seen along roadsides following summer or fall rain.

The trumpet-shaped flower can be pink, bluish, purple, or rarely white. The corolla tube is 1 to 1½ inches long and about as wide. The fruit is an ovoid capsule about 4 mm wide by 6 mm long. Flowering occurs June through November following summer monsoons or fall rain.

The bird's foot-shaped leaf blades are palmately divided into three to five slender lobes that are 1 to 3 mm wide by 2 to 8 cm long. The basal pair of lobes is usually divided into another three lobes. The dense stems and leaves give the plant a compact growth form.

Bird's foot Morning Glory growing with Summer Poppy along Mexico Highway 15 near Benjamin Hill.

WHERE TO SEE IT
Growing in the gravel shoulders of Mexico Highway. 15 between Nogales and Hermosillo following summer or fall rain. Easily recognized by the bright purple flowers and slender bird's foot leaves.

38

Area B

Dunes, Estuaries, and Coastal Wetlands

B-1 COASTAL SAND VERBENA, Red Sand Verbena, Alfambrilla

Scientific name: *Abronia maritima*
Family name: Nyctaginaceae. Four o'clock family

Coastal Sand Verbena is a colorful dune succulent forb with sprawling or erect stems, to 12 inches tall, growing above the tide line on coastal dunes. Numerous long trailing stems form large mats 5 to 10 feet across, entirely covering the sand.

The small flowers are dark crimson to red-purple and grow in clusters. Coastal Sand Verbena can flower any time of year, especially following rain.

The plant has fleshy leaves and stems. The leaves are round or oblong and 1½ inches across.

Coastal Sand Verbena resembles Sea Purslane (B-2), another dune succulent, often growing on the same sites. Sea Purslane can be distinguished by its narrow, succulent leaves versus the rounded leaves of Coastal Sand Verbena.

Both are important dune stabilizers.

Off-road vehicle use has destroyed most Coastal Sand Verbena and Sea Purslane mats along the beach at Kino Bay.

Green strips of Coastal Sand Verbena. Pelican Island is in the upper left, with Tiburón Island in the background and Kino Bay on the right.

WHERE TO SEE IT
Common on the dunes at Kino Bay and the tidal flats behind the townsite.

40

B-2 SEA PURSLANE, Cencilla

Scientific name: *Sesuvium portulacastrum*
Family name: Aizoaceae. Carpetweed family

Sea Purslane is a low (to 12 inches tall), perennial, succulent forb with runners forming large mats. It grows on coastal dunes above the tide line, near estuaries and on coastal wetlands. It closely resembles and is related to Western Sea Purslane *(S. verrucosum)*, which grows in many of the same sites.

The small, pretty, pink flowers can appear from March through November. The flowers have sepals but no petals. The sepals are succulent, green outside, and pink inside. The succulent leaves are narrow, flattened, and dense. Leaves and stems are often red-green.

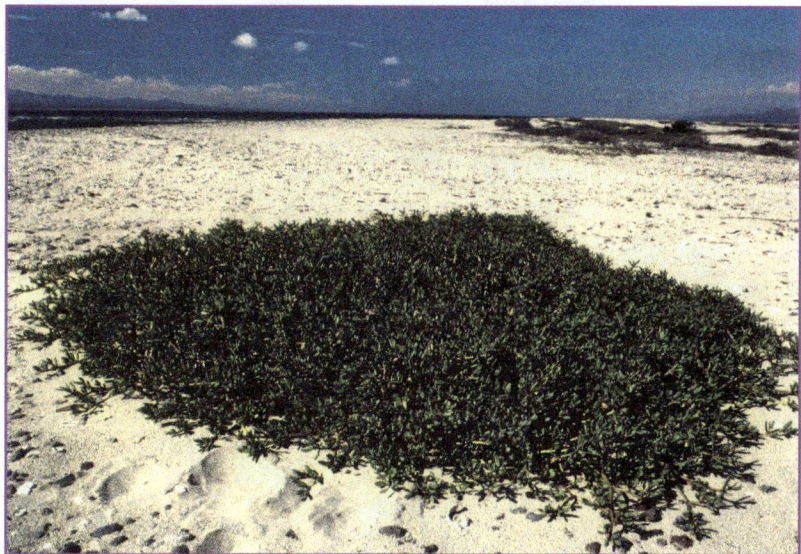

The mat-forming characteristic of Sea Purslane resembles Coastal Sand Verbena which is also common on coastal dunes. The two can be distinguished by their leaf shapes. Purslane leaves are narrow, and verbena leaves are rounded.

Seri Indians considered Sea Purslane to be a "soft" plant. It was used to line the inside of a turtle carapace or a basket to provide a bed on which to place meat to keep it clean.

Sea Purslane can be grown as an interesting house plant. Place several cut stems in a vase or glass of water. Add a half stick of crushed Job's Spike fertilizer and place in a window. The plant quickly develops roots, grows at a modest rate, and will eventually bloom.

WHERE TO SEE IT
Off-road vehicle use has probably destroyed most plants along the coast around Kino Bay, but some can be seen in protected areas near homes along the beach. Some plants grow on the dunes of vacant lots on the north side of Mar de Cortés Boulevard. Sea Purslane, Coastal Sand Verbena, and other dune plants are much more common on lightly-used beaches southeast of Kino Bay.

B-3 PALMER ALKALI HEATH, Hierba Reuma, Saladito

Scientific name: *Frankenia palmeri*
Family name: Frankeniaceae. Frankenia family

A short (1 to 2 feet tall) rounded shrub with dense, brittle stems. The evergreen leaves are grayish, 2 to 4 mm long, and nearly round in cross section.

   Tiny white to pinkish flowers appear November through May, but shrubs may flower at almost any time of year. The fruit is a small, ovoid capsule.

   Palmer Alkali Heath grows in nearly pure stands on salty soils behind coastal dunes and around the margins of estuaries and playas. It is extremely salt tolerant, taking up salt through the roots and secreting it through the leaves.

A nearly pure stand of Palmer Alkali Heath near Estero Santa Rosa, northwest of Kino Bay. Dos Palmas and Western Horizons R.V. Court are at the foot of the mountain in the upper left corner of the picture.

WHERE TO SEE IT
Coastal wetlands (salt flats) directly behind (north) of New Kino and along the road to the Kino Bay airport. Also in coastal wetlands along the highway entering Kino Bay where it is growing in association with Desert Saltbush (*Atriplex polycarpa*). These wetlands are dry for about half of the year.

B-4 SAND CROTON, Hierba del Pescado

Scientific name: *Croton californicus*
Family name: Euphorbiaceae. Spurge family

Sand Croton is a low shrub (to 3 feet tall) that is densely-branched and grows in small clumps. It is a common plant on coastal dunes above the tide line. It often grows in association with Coastal Sand Verbena and Sea Purslane where the three play a roll in dune stabilization. Off-road vehicle use and human traffic have eliminated most of these three species along the beaches at Kino Bay.

The shrubs are easily recognized by their silvery leaves which are very attractive. The leaves are 1 to 4 inches long, 2 to 8 mm wide. and densely pubescent. The plants have tiny male and female flowers. Male flowers appear on elongated racemes. The flowers are not attractive.

Sand Croton flowers at various times from February through October.

The leaves have a strong odor. Seri Indians crushed them to release a toxin; then threw them into estuaries to kill fish, which may account for the Spanish name – Hierba del Pescado (herb of the fish). Many species of plants in the spurge family contain chemicals that are caustic or poisonous. See Ashy Limberbush (C-23).

Sand Croton (silver-gray) growing with Sea Purslane (green) on a beach southeast of Kino Bay.

Silver-gray leaves of Sand Croton

WHERE TO SEE IT
Common on the dunes at Kino Bay and dunes on vacant lots along the north side of Mar de Cortés Boulevard in New Kino.

43

B-5 RED MANGROVE, Mangle Dulce, Mangle Colorado, Mangle Tinto, Candelon, Mangle Salado, Mangle Rojo.

Scientific name: *Rhizophora mangle*
Family name: Rhizophoraceae. Red mangrove family

Red Mangrove is a shrub or small tree (to 20 feet tall), growing in impenetrable stands on muddy-sandy soil of the intertidal, shallow sea water of estuaries. They have stilt roots called aerial prop roots that provide stabilization and oxygen. These aerial roots can develop from any tree branch and they grow rapidly.

Leaves are thick, leathery, ovate or broadly elliptic, and 3 to 6 inches long. They occur in groups at the ends of branches. The flowers are small, whitish to yellowish-green, axillary with the leaves, and appear March through November. Flowers have four fat petals. The fruiting pod, called a "propagule," is green, cigar-shaped, and about 8 to 10 inches long. The actual fruit portion of the propagule is only about 1 inch long. Even before it drops from the tree, the "radical" (first root) bursts from the fruit and extends the length of the propagule to 8 to 10 inches. The propagules fall into the sea and may float for miles before washing up on a beach. They are commonly seen at Kino Bay and all along the coast. Propagules can live up to a year or more without rooting.

In the Kino Bay area, Red Mangrove typically grows with Black Mangrove (*Avicennia germinans*) and White Mangrove (*Laguncularia racemosa*).

HOW TO TELL THEM APART:

1. Red Mangrove has shiny green, semi-succulent leaves, 3 to 6 inches long, with no pubescence. Leaves are in whorls. It is the only one with stilt roots. No pneumatophore roots (definition under Black Mangrove, below).

2. Black Mangrove has a light gray-colored leaf, 2 to 5 inches long, not shiny, with dense scurfy pubescence beneath. The leaves are opposite, not in whorls. Leaves of Red and White Mangroves have no pubescence. Both Black and White Mangroves have pneumatophore roots resembling drinking straws that protrude above water to get oxygen for subterranean roots.

3. White Mangrove has the smallest leaf of the three (to 3 inches long). The leaf is not shiny. There is a conspicuous pair of glands on the petiole near the leaf blade that secrete a sweet solution. Pneumatophore roots are thick, knobby, and branched, often developing from the shallow horizontal root system.

44

Mangrove forests appear to require, or at least benefit from, inland fresh water run-off that brings minerals, specifically nitrogen, phosphorous, and iron, elements in which seawater is deficient.

Mangrove trees are valued for building material, as the wood is resistant to water. Seri Indians ate the fruit and made a black dye from the roots. A Seri Indian family at Punta Chueca grows and sells nursery stock of all three species.

Mangrove forests are among the most productive and biologically complex ecosystems on earth. Birds roost in their canopy, shellfish attach to the roots, and they provide nursery grounds for fish, a source of food for some animals, and a source of nectar for bats and honeybees. They catch sediment from run-off to become soil-builders, and they buffer the erosive power of waves.

Mangroves are widespread throughout the tropical world where they are the supermarkets, lumberyards, fuel depots, and pharmacies of the poor. Mangroves are highly efficient carbon sinks. They absorb large amounts of carbon dioxide and reduce greenhouse gas. Despite these qualities, mangroves are in decline. One of the greatest threats is from commercial shrimp farms which have destroyed thousands of acres of mangrove forests all over the world. (From "Forests of the Tide," by Kennedy Warne and Tim Laman, *National Geographic*, February 2007).

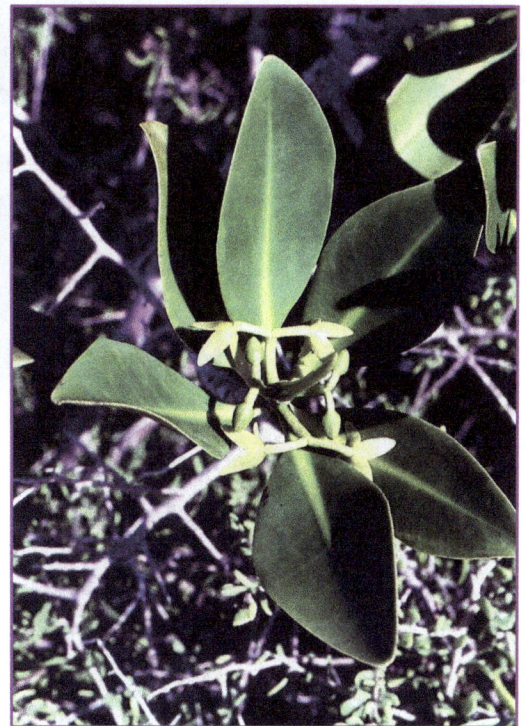

WHERE TO SEE IT
Estero Santa Rosa 2 miles south of Punta Chueca and Estero Santa Cruz at Kino Bay.

(Top) Cigar-shaped propogules or seedlings of Red Mangrove washed up on a beach. (Above and center) Flowers of Red Mangrove on two T-shaped flower stalks.

45

B-6 PICKLEWEED, Glasswort, Samphire

Scientific name: *Salicornia bigelovii*
Family name: Chenopodiaceae. Goosefoot family

Pickleweed is an annual forb growing in rather dense stands in salt marshes and in estuaries between the mangrove trees and the sea. It is often submerged with tidal fluctuations. It is classed as a halophyte because it is a salt tolerant plant.

Flowers are in tight clusters along the top of the stems. Flowering often occurs from May through June, with seeds maturing September through October. Seeds germinate November through January and grow rapidly.

The stems are modified leaf tissue. They are bright yellow-green, succulent, glassy, and few to moderately branched.

There has been a great amount of research devoted to Pickleweed in various parts of the world, including Mexico. Selections of the plant have been developed to grow as a sea-water irrigated crop. The seed contains from 26 to 33% oil and 31% protein. It is a potential source of feed for livestock and chickens. In some parts of the world it has been planted to rehabilitate degraded coastal lands. Pickleweed appears to be a potentially valuable crop for subtropical coastal deserts.

Pickleweed at Estero Santa Rosa

Illustration at left shows dense seed heads at the tops of branches of Pickleweed.
(Illustration from *An Illustrated Flora of the Northern United States, Canada and British Possessions*, Vol. 2, 1913, by Britton and Brown.)

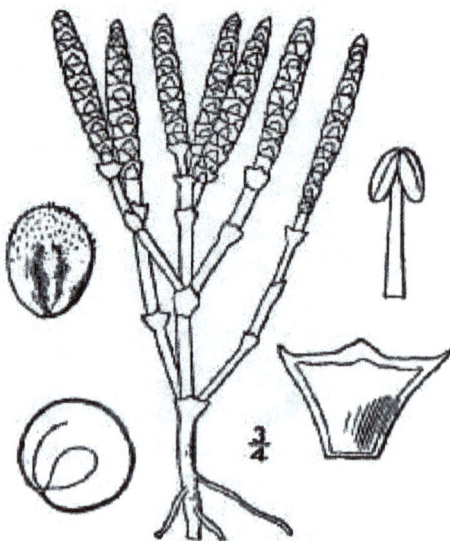

WHERE TO SEE IT
Pickleweed and other halophytes (salt-enduring plants) are being grown commercially at the Seawater Foundation Farm in New Kino. Wild plants can be seen at Estero Santa Rosa northwest of Kino Bay.

B-7 IODINE BUSH, Chamizo
Scientific name: *Allenrolfea occidentalis*
Family: Chenopodiaceae. Goosefoot family

A perennial, densely-branched, succulent, scraggly shrub, 2 to 5 feet tall; often broader than tall, growing in salty soil and near estuaries. Iodine bush is a halophyte, or salt-enduring plant.

Flowers are tiny, lacking petals, and growing in dense terminal and lateral spikes. Flowering occurs July through September, with seed ripe in mid-winter. The seeds are ovoid, about 1 mm long, reddish-brown, smooth, and shining. Seri Indians gathered the seeds, toasted and ground them, and made a gruel to eat.

Leaves are triangular, fleshy, 2 mm long, and falling early. The plant appears leafless most of the year.

The older branches are woody. Younger branches are fleshy, bead-like, and green to red-orange. When stressed by drought, the plants appear almost black.

Iodine Bush has a high water requirement. It grows near estuaries, behind dunes, and on saline flats usually close to the sea. It is most common on saline sites with a perennial high water table but does grow on adjacent sites lacking this moisture advantage. During periods of drought, it will almost disappear from the latter sites but reestablishes quickly when rainfall is abundant. Iodine Bush can form rather dense stands.

The young, succulent stems of Iodine Bush resemble Pickleweed *(Salicornia bigelovii),* and the two halophytes are commonly found growing together.

WHERE TO SEE IT
Iodine Bush is common along the road in the salt flat directly behind New Kino where it is growing with other halophytes including pickleweed. It is grown commercially at the Seawater Foundation farm in the same area.

HOW TO TELL THEM APART:

Pickleweed is a short, annual forb, 8 to 18 inches tall and not woody. Iodine Bush is a perennial shrub, 2 to 5 feet tall, with woody older stems and succulent younger stems.

47

B-8 GOLONDRINA, Sand Mat, Spurge

Scientific name: *Euphorbia* spp.
Family name: Euphorbiaceae. Spurge family

A prostrate, mat-forming herb growing on dunes, disturbed areas, and a wide variety of sites in the Gulf area. A common, somewhat weedy spurge. There are several species in our area and all look much alike. The stems have white milky juice. Some Golondrina species are annuals and others are short-lived perennials.

The tiny white flowers are numerous. Golondrinas may grow and flower any time of year following rains. Stems, leaves, and flowers disappear during extended dry periods.

What appears to be the flower is actually an inflorescence called a cyathium. The cyathium consists of male and female flowers congested inside a cup. The white petal-like things around the edge of the cup are called appendages and are not petals. The cyathium type of inflorescence is characteristic for all species within the Spurge family.

Golondrina growing with Coastal Sand Verbena on an inland dune at Kino Bay.

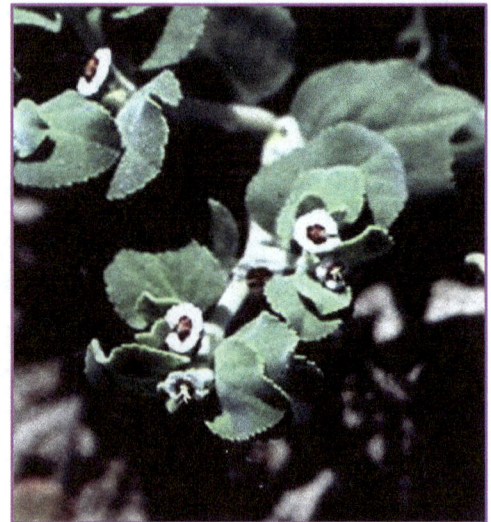

WHERE TO SEE IT
Golondrinas can be found on coastal dunes, inland dunes, roadsides, vacant lots in the Kino Bay town site, and nearly every other habitat type in our area. A very common plant.

Area C

Kino Bay and Vicinity

C-1 TRIXIS, Plumilla
Scientific name: *Trixis californica*
Family name: Asteraceae (Compositae). Sunflower family

A dense shrub, 1 to 3 feet tall, growing alone or under another bush. giving the appearance that its numerous, small yellow flowers belong to the host. Leaves and flowers appear any time after rains. The dense heads of bright yellow flowers are about ¾ inch wide. The peak blooming periods are from February through June and October through November.

Leaves are from 1 to 3 inches long, lance-shaped, and upright. Leaves wither during dry or cold periods. The stems are often covered with dense brown hairs.

The Spanish name, Plumilla, means "little feather" or "little plume" for the golden bristles on the seed-like fruit.

Seri women drank a tea made from the leaves to hasten childbirth. Leaves were also smoked like tobacco.

Trixis is widely distributed in southern California, southern Arizona, most of the Baja peninsula, and in Sonora, Mexico. It is only an occasional plant here at Kino Bay. It grows along washes, on rocky slopes, bajadas, and gentle slopes in association with elephant trees and limberbush species. It often grows under trees. Trixis is easily missed during dry periods when the leaves and bright yellow flowers are not apparent. Look for it after moderate rainfall.

WHERE TO SEE IT
Desert lots behind Kunkaak R.V. Court in Kino Bay and brushy areas on arroyo terraces northwest of Kino Bay. Not common in the area covered by this book. Trixis is seldom noticed until its bright yellow flowers appear.

C-2 BALLOON VINE, Tronador

Scientific name: *Cardiospermum corindum*
Family name: Sapindaceae. Soaptree family

A perennial woody vine, scrambling on the ground or climbing into a host shrub or tree as high as 25 feet. The dense foliage sometimes nearly covers the host plant.

Balloon Vine flowers are small, numerous, lightly fragrant, and whitish to pinkish, with four sepals and four petals. Flowers bloom year-long. The brown, papery pods resemble Jack-o-lanterns and persist on the vines long after they dry. The pods are abundant, about 1 to 2 inches across, and easily seen from a distance. They make attractive floral decorations. Seeds are black with a white "eye."

Leaves are pinnate with numerous attractive leaflets. The leaflets are 5 to 20 mm long, serrate, and delicate-appearing.

Leaves and flowers wilt during dry periods and are quickly replaced following rains any time of year.

WHERE TO SEE IT
Rather common on sites supporting mid-to-tall shrubs immediately northwest of Kino Bay on bajadas and along arroyos between the mountains and the sea. Uncommon southeast of Kino Bay. Rather common along Mexican Highway 15 just north of Hermosillo.

C-3 WOLFBERRY, Tomatillo (Little tomato)

Scientific name: *Lycium* spp.
Family: Solanaceae. Nightshade family

There are four species of Wolfberry in the Kino Bay vicinity and all are somewhat similar. They are compact, many-branched shrubs, often forming thickets. Wolfberries are dense shrubs, 4 to 7 feet tall, with small stems branching at 90 degree angles from larger stems. This right-angle branching is especially obvious when the shrubs are leafless. Plants are thornless, but the short side branches have sharp tips. Photos are of Fremont Wolfberry.

Leaves are short, more or less succulent, shed in dry periods, and regrown following rains. Winter rains produce flowers in February and fruit in March and April; however, flower and fruit production may occur any time of year following rains. The numerous tiny flowers are ¼ to ½ inch wide (see colors for different species below). Some plants have flowers with four petals, some with five, and some may have both.

Berries of all but *L. californicum* were relished by the Seri Indians, who also used the hard wood of the stems to make bows and fore shafts for arrows. Of the many kinds of wood used for making bows, the Seri preferred Wolfberry.

THE FOUR SPECIES OF WOLFBERRY IN THE VICINITY OF KINO BAY:

1. *L. fremontii:* **Fremont wolfberry** (all photos). Flowers pale to dark lavender, with four or five petals. Each plant commonly produces large amounts of juicy, bright red or red-orange, fleshy, round or oblong berries about 12 mm long (range 7 to 18 mm). Berries from this plant were the most extensively used of all four species by the Seri Indians.

52

It is widespread in desert lowlands. It is readily recognized by the glandular hairs on the leaf and on the pedicel and calyx of the flower. Other distinguishing features are the long pedicels, often large flowers, and non-exerted stamens. Flowers February through May and often August through December.

2. *L. californicum:* **Wolfberry or Frutilla.** Flowers white, with purple tinge. Berry scarlet, scarcely fleshy, smaller than the other three species. The only Wolfberry whose fruit is bitter and not eaten. It is the only species with leaves round in cross section. Other species are semi-succulent but not round in cross section. Twigs are thicker than the other three species. Common near beaches. Flowering February through July.

3. *L. andersonii:* **Desert wolfberry.** Flowers greenish-white with lavender lobes. Corolla tube funnel-shaped, mostly very narrow and nearly tubular. Flower petals four or five. The fifth petal sometimes larger than the others. Fruit bright orange, fleshy. Flowering February and March.

4. *L. brevipes:* Flowers white to violet. Fruit red, fleshy. Flowering February through April.

Wolfberry is sometimes grown as a landscape plant. Plants can be grown from cuttings and require minimal water after establishment.

WHERE TO SEE IT
Fremont Wolfberry grows along the dirt road ½ mile due north of the Pemex station in Old Kino. It also grows on dunes on vacant lots on the north side of Mar de Cortés Boulevard in New Kino. *L. californicum* **grows on the leeward side of beach dunes around Kino Bay.**

53

C-4 WHITE-BARK ACACIA, Palo Blanco.
Scientific name: *Acacia willardiana*
Family: Fabaceae (Leguminosae). Legume family
Subfamily: Mimosoideae. Mimosa subfamily.

Perhaps the most graceful tree in the Sonoran Desert. A thin, wispy, short tree, to 25 feet tall, with whitish-yellow, exfoliating bark. Grows on rocky slopes, cliffs, and in arroyos near mountains. Tiny yellow catkin-like flowers appear in February, and the tree continues to flower into July. Flowers are followed by rather large, thin, flat, brown pods.

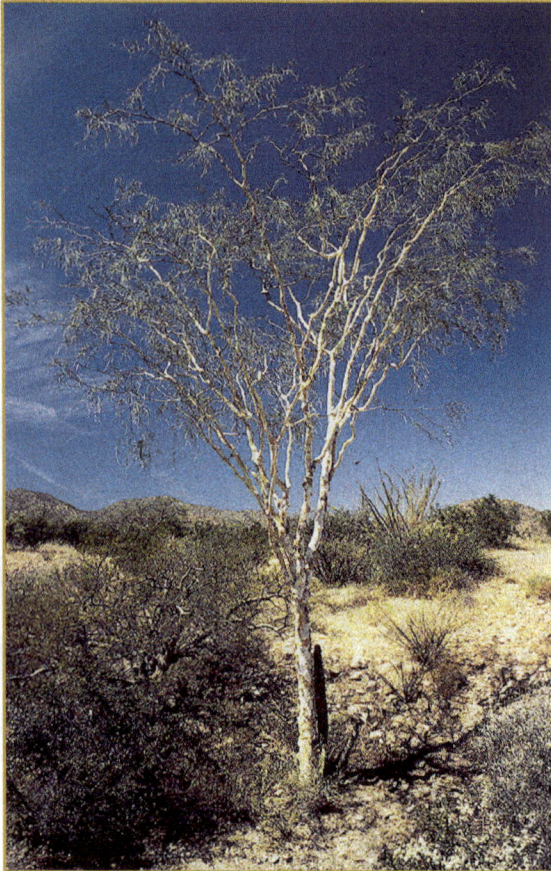

White-bark Acacia is the only Sonoran Desert plant, and the only American acacia, with "phyllodic" petioles called "cladophylls" which are leaf-bearing petioles that manufacture chlorophyll. This feature is rare in the plant world since leaf petioles of most plants do not manufacture chlorophyll. The long (6 to 12+ inches) cladophylls produce tiny leaflets that are short-lived and drop off, leaving the graceful cladlophylls to continue providing nourishment for the tree.

White-bark Acacia is endemic to Sonora and to Tiburón and San Pedro Nolasco Islands. Its range on the mainland includes the coastal mountains from the Sierra Seri (behind Kino Bay), south to Guaymas. It is also found near Alamos and on the Rio Yaqui near Soyopa in extreme southern Sonora. It is occasionally planted as a landscape plant in parts of Arizona and Sonora, where temperatures do not drop below 42 degrees F.

Seri Indians used the hard wood for bows, carrying yokes, foreshafts for harpoons, chisels, pry bars, and digging sticks. Bows were also made from Wolfberry (preferred), Catclaw, *Caesalinina,* and Coursetia.

Flowers and cladophylls

WHERE TO SEE IT
From Kino Bay R.V. Park in New Kino, drive north past the shrimp germ plant and over the hill into a large desert valley. At the first junction, follow the road to the right (north) that leads to a large arroyo. Follow the road east up the bottom of the arroyo. Trees are scattered along the sides of the arroyo. This road eventually joins the north-south road to Punta Chueca. Other trees can be seen on a rocky slope on the west side of Mexican Highway 15 just north of where the highway divides to San Carlos and Guaymas.

54

C-5 OCOTILLO, Common Desert Ocotillo, Coach Whip

Scientific name: *Fouquieria splendens*
Family: Fouquieriaceae. Ocotillo family

A small wand-like woody shrub to 20 feet tall, with many slender stems originating from a woody base near the ground. Flowers are red, tubular, about 1 inch long, and in clusters along a spike about 10 inches long, resembling a flag at the tip of the stems. In our area, Ocotillo may flower any month following rain, but the typical flowering period is from February through June. Leaves appear following rains and drop off during dry periods. Ocotillo is not a cactus.

The tan upper portion of the stem is the long shoot with long shoot leaves of the current year. The leaf petioles branching at right angles to the stem will eventually become new thorns.

There are four species of *Fouquieria* in a 70-mile radius of Kino Bay. They include Tree Ocotillo, Adam's Tree, Boojum, and common Ocotillo. All are included in this book.

HOW TO TELL THEM APART:

In the absence of flowers and leaves it is difficult to tell Tree Ocotillo from Adam's Tree. However, common Ocotillo is easily differentiated from these two by its very short trunk and thin stems. Mature Tree ocotillo and Adam's Tree have definite short trunks and at least several limbs thick enough (3 -8 inches) that they are used for fence posts.

Photos on the following page show how to tell the three by floral differences.

Mexicans harvest the stems and plant them to make living fences and corrals. Stems cut too close to the growing point causes the cut stump to die. Leaving a longer stump allows for regrowth and assures perpetual harvest. Some newly-planted stems (cuttings) leaf out in a short time, while others may not produce leaves for several years. All cuttings usually produce flowers every year. Ocotillo is a common landscape plant.

The historic use of Ocotillo stems for making shelter frames is still practiced by the Seri Indians. Shelters can be seen at the Seri village of Punta Chueca. The

THE FLORAL DIFFERENCES OF THREE SPECIES OF OCOTILLO

Common Ocotillo:
Flower pedicels short (4 to 8 mm).
Flowers up right on a long, flag-like spike.

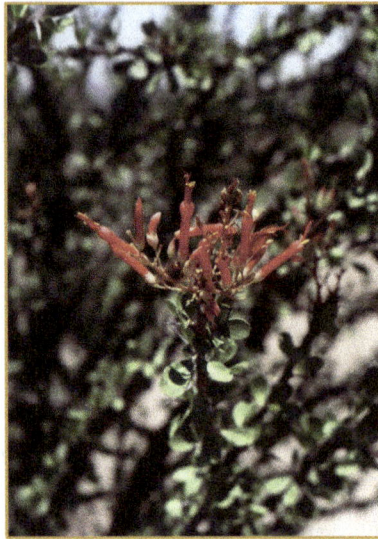

Adam's Tree:
Flower pedicels short (2 to 6 mm).
Flowers upright and grouped in a
cone shape.

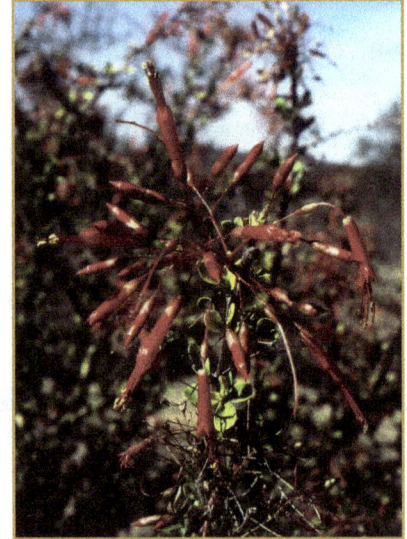

Tree Ocotillo:
Flower pedicels very long (to 30
mm). Flowers drooping.

large ends of stems are stuck in the ground to form a square or rectangle. The tops are bent over and tied to stems on the opposite side. Brush, driftwood, grass or most anything handy is attached to make walls and a roof. Many of these stems rooted and grew, showing the outline of the shelters long after they have been abandoned.

SHORT SHOOT AND LONG SHOOT STEMS AND LEAVES OF OCOTILLO

All species of the genus have what are called short shoot and long shoot leaves and stems. The current annual growth at the end of the stem is called the "long shoot." (See photo on previous page.) As the long shoot elongates, it produces leaves with stout petioles. When the leaf blade breaks off from the petiole, most of the petiole and the midrib of the leaf become woody and form a spine. These leaves are called "long shoot leaves" because they are produced on the long shoot.

Leaves that sprout from the older part of the stem (below the long shoot) appear above the base of spines from a growing point resembling a wart. This point is called a "short shoot" because it only grows a couple millimeters during the life of the plant. The leaves are called "short shoot leaves." Short shoot leaves drop off when conditions turn dry, and new leaves appear following rains. However, long shoot leaves are produced only in summer when the stalks are elongating.

WHERE TO SEE IT
Ocotillo has a large range in the Sonoran Desert. It is most abundant on limestone soils and on sandy or rocky soils on a rise that is well-drained. It is common on sandy rises just east of Kino Bay along State Highway 100.

C-6 PINICUA, San Juanico

Scientific name: *Jacquinia macrocarpa* var. *pungens* (*J. pungens*).
Family: Theophrastaceae. Theophrasta family

A small evergreen tree to 20 feet tall, with smooth, gray bark and thick trunk and stems. Thornless and densely-branching. An occasional tree most often seen growing singly or in small groups on saline flats well back from coastal dunes. Some trees have dense, well-rounded crowns, while others have grotesque shapes, especially when decadent, as in the photo at lower right.

Pinicua flowers are small (⅓ inch wide), reddish-orange, and rigid. They are attractive and have the appearance of dried flowers. The flowers appear in spring, usually in May. The fruit is a round, tan pod about ¾ inch across. The seeds are eagerly sought by birds and by rodents who leave little piles of broken pods where they have collected them to eat the seeds.

Stems end in sharp terminal spines. Leaves are narrow (1 to 2 inches long and ¼ inch wide), dark green, thick, sharp-pointed, and with edges rolled inward.

Seri Indian girls made necklaces of the flowers. Mexicans carve pictures and names in the smooth gray bark.

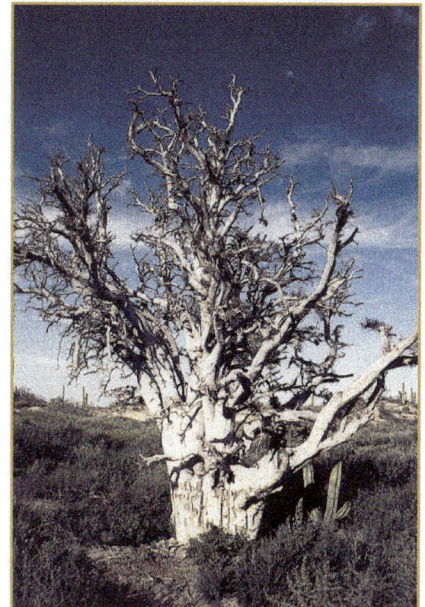

WHERE TO SEE IT

Pinicua trees are growing in salt flats along the dirt road to Punta Baja, southeast of Kino Bay. A few trees can be seen west of Calle 36 Norte between 4 and 10 kilometers north of the junction with the highway to Kino Bay. Pinicua is not a common tree, but is widely dispersed around Kino Bay.

C-7 ICE PLANT, Flor del Sol

Scientific name: *Mesembryanthemum crystallinum*
Family: Aizoaceae. Aizoon family

Ice Plant is a low, succulent, annual or perennial herb growing singly or in mats of up to 20 inches across and usually only a few inches high. It is profusely branched. The herbage is covered with transparent, glistening vesicles that look like heavy dew – thus the name "Ice plant."

Ice Plant is a native of South Africa that was introduced into the United States as a landscape plant. It escaped cultivation and has since naturalized in sandy soils along the coasts in California, Baja California, and Sonora.

Flowers are very striking in appearance. They are about 1 inch across, with whitish to reddish petals. Flowering occurs February through June. The leaves are mostly rounded, from ⅓ to 2 inches long, undulating, and succulent.

Ice Plant is a popular garden plant in warm climates. Be advised! It shares its name with several other succulent ground cover plants that are also landscape favorites. Plants within the *Mesembryanthemum* genus are simply referred to as "mesems" by gardeners.

WHERE TO SEE IT
Ice Plant can be seen growing along the airport road near the large city water tank just north of New Kino. In this area, it appears to be an annual that only grows when there is sufficient rainfall. If winter-spring rains are abundant, Ice plants are also abundant, especially on salt flats well back from coastal dunes. It is nearly absent in dry years.

C-8 CREOSOTEBUSH, Hediondilla (little stinker), Gobernadora

Scientific name: *Larrea tridentata (L. divaricata)* ssp. *tridentata*
Family: Zygophyllaceae. Caltrop family

🌿 *Nurse plant. One of the most important plants to the ecology of native deserts.*

Creosotebush is a semi-evergreen shrub, 3 to 6 feet tall, with thin, gray stems and open growth. Creosotebush is a "facultative evergreen," which means it can carry on leaf functions during hot or cold months. It has a tap root and many shallow lateral roots to maximize water uptake and remain green. It is often the dominant species in the driest areas and on abandoned farmland. It is widespread in dry areas of northern Mexico and southwestern United States.

Creosotebush has a profusion of small yellow flowers from February through April, but it may flower any time following rain.

Creosotebush prefers well-drained sandy or rocky soils and almost never grows where there is a shallow, restrictive clay layer. It is often abundant on limestone soils because moisture is retained in the cracks of the bedrock, giving the plant a longer season of water availability. In contrast to succulents such as Saguaro cactus that can take up water only from saturated soils, Creosotebush can obtain water from soil that feels dust-dry to the touch. (Refer to the note with Saguaro, C-9.)

Creosotebush is considered the most drought-tolerant plant in North America. A study in northeastern Baja California revealed that Creosotebush died back to the ground (root crown) after four years with no rain, then recovered following rains in the fifth year.

Creosotebush is one of the world's longest-lived plants. Although each stem only lives one to two hundred years, new plants produced from the outer edge of the root crown expand the plant into a gradually extending circle of plants. Over centuries the circle breaks down into groups of individual plants. One such circle in the Mojave Desert is 26 feet in diameter and several thousand years old. Where you see large circles (clones) of Creosotebush, it indicates the presence of very old plants and an older, stable soil environment. Studies show some of these circles could be as much as 11,700 years old. Creosotebush circles are not common around Kino Bay.

Creosotebush has resins on and in its leaves, which make it unpalatable to livestock and wildlife. Rabbits have been known to eat the leaves under emergency conditions in the absence of other vegetation. The resin is responsible for the leaf's shiny coating. The major resin is dihydroguairetic acid (NDGA), one of the best preservatives known. Until synthesized in 1967, it was used to protect frozen food, especially meat, and to retard rancidity in lard and vegetable oil. The resin gives off an odor after rainstorms that smells like creosote. Mexicans call the plant Hediondilla, meaning "little stinker."

Creosotebush is occasionally used in natural landscaping where people enjoy the brilliant leaves, profuse yellow flowers, and the aroma following rain. Pruning encourages a thicker canopy, and irrigation encourages the plant to retain its leaves and to flower year-round.

Creosotebush branches are sometimes encrusted with a reddish-brown substance called "lac," caused by the excretions of *Tarcardiella larrea,* a scale insect. The Seri Indians melted lac and formed it into a ball for future use as a glue. When re-heated and applied as glue, it dried and formed a strong bond. It was used to repair pots, seal lids of storage jars, and for hafting harpoons and arrow heads. Lac was used to paint a series of three circles near the end of the main shaft of an arrow. These black rings identified the arrow as Seri. Indians of other tribes made a strong concoction from the leaves and stems to treat arthritis. Leaves were also used as an antiseptic and styptic to soothe cuts and bruises.

Creosotebush is a pioneer species, establishing on abandoned farm land and heavily overgrazed or physically disturbed native range where it often becomes the major plant component if grazing is left unchecked. With controlled grazing, Creosotebush begins to fulfill its roll as a nurse crop that encourages establishment of other vegetation.

WHERE TO SEE IT
Common on abandoned farm land along Calle 36 Norte, beginning just north of the junction with the highway to Kino Bay.

C-9 SAGUARO, Sahuaro

Scientific name: *Carnegiea gigantea*
Family: Cactaceae. Cactus family

A tall columnar cactus found only in the Sonoran Desert. It is common around Kino Bay where it often grows with Cardón cactus (C-10), which it resembles. See **HOW TO TELL THEM APART** under Cardón cactus. The genus name honors the Carnegie family who established the former Carnegie Desert Laboratory near Tucson in 1902 to study desert plants.

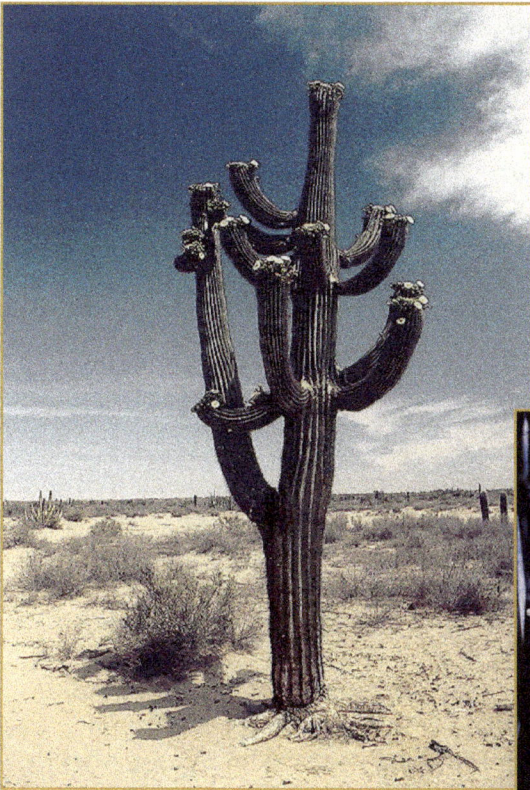

Saguaro is the largest cactus in the United States. In 1991, the tallest in Arizona measured 59 feet. It has since died, and the current (2004) record nholder is 51 feet tall, found by Mike Hallen near Florence, Arizona (*All About Saguaros*, 2008, by Leo F. Banks). The largest columnar cactus in the world is the Cardón, which does not grow in the United States.

During the fore-summer period of April through June when the desert is hot and dry, most plants are dormant, awaiting summer rains. Saguaros use their stored water to flower and set fruit at this time. Saguaros flower every year but produce more flowers following winters with heavy rainfall. Large (3 inch) white flowers are produced from April through June. Flowers usually appear as a halo around the top of the stem and arms, but some Saguaros will

have flowers extending down the stem a foot or more.

Flowers have an aroma like an over-ripe melon. The fleshy fruits turn red or purple at maturity, usually in July, and split open, revealing a crimson red interior and many tiny black seeds. Seeds and fruits are eagerly sought by birds and rodents. In some years of heavy rainfall, Saguaros may flower as late as September and October.

Bats, bees, and white-winged doves are among the pollinators of Saguaro flowers. After wintering in tropical Mexico, lesser long-nosed bats migrate up the Sonoran coast in the spring, feeding mostly on the

Ripe fruit burst open.

flowers of several species of columnar cacti which supply their complete diet during their northern migration. These bats summer in Arizona and return to tropical Mexico in the fall. On the return route, they feed on and pollinate the flowers of several species of agaves from July

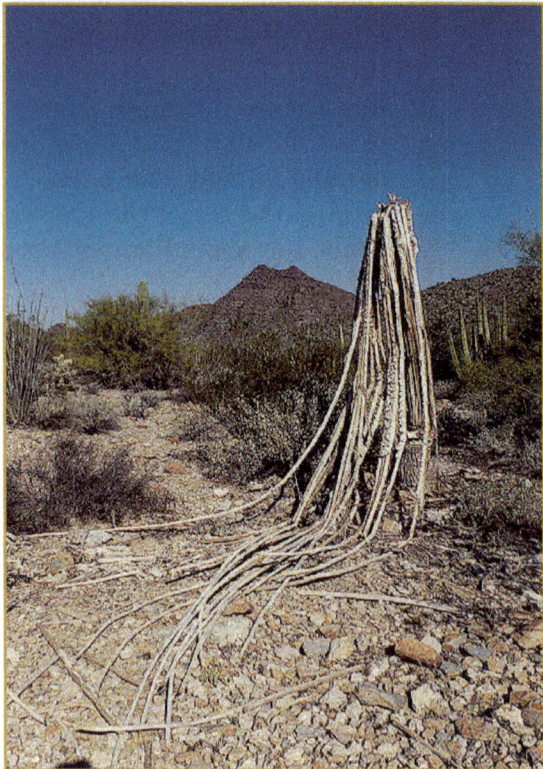

through September. A study at Kino Bay showed that bats appear to be minor players in Saguaro pollination.

Although Saguaros grow on a variety of sites, the densest stands are on rocky hillsides with well-drained soils. Their distribution in the Sonora Desert shows they favor the eastern portion with its higher proportion of summer-to-winter rain. Saguaros are totally absent on the Baja peninsula, in part because of a predominance of winter rain in the northern half.

The root system of Saguaro cactus consists of a shallow tap root, 2 feet deep, and many long lateral roots a few inches beneath the soil surface that radiate out as far as the tree is tall. In a process called "passive diffusion," Saguaro roots can take up water only when the soil is wetter than their own moist interiors. This water uptake process is common to stem succulents such as Saguaro and barrel cacti. Research shows that a fully-hydrated, mature Saguaro is more than 90% water and can loose up to 81% without serious harm. Saguaro seedlings can survive losing up to 55% of their stored water. Its thirteen to twenty woody ribs allow the stem to expand like an accordion as water is added to the central core. Through a process called "thermal inertia," heat is absorbed during the day and released at night, which also aids in preventing the tissues from freezing.

Annual height growth depends on age and location. During the first few years, a Saguaro may grow only 1 to 2 mm PERyear. A study at Saguaro National Park East showed, as a rule of thumb, a Saguaro should reach 12 feet in fifty years, 22 feet in seventy-five years, and 30 feet in one hundred years. The average heights of fifty-year-old Saguaros at two other locations in Arizona were: 3 feet at Organ Pipe National Monument, and 7 feet at Saguaro National Park West. Saguaros grow rapidly with irrigation, but over-watering can be fatal. On average, Saguaros may produce a few flowers and fruits at thirty to thirty-five years of age. When seventy-five years old, one can produce as many as one hundred flowers and fruits per year.

Another study showed Saguaros have most arms on the south side of the main trunk. Ribs on south-facing sides are about 20% closer than those on the north side. This increases shading and reduces plant temperature, the same as for barrel cactus.

Many seedlings establish near rocks or beneath nurse plants such as Ironwood and mesquite. They grow slowly and depend on the protective cover of the host tree to provide shade and prevent damage from ungulates. Large scale harvest of mesquite and Ironwood, followed by heavy soil erosion and trampling damage by livestock, has

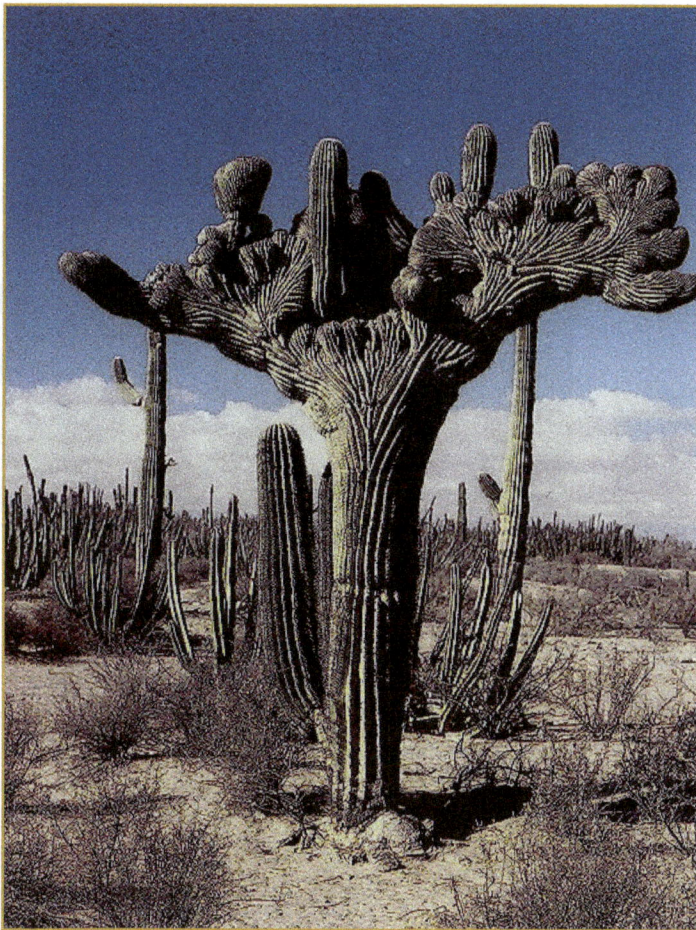

A very large crest on a Saguaro near Calle 20 Sur southeast of Kino Bay.

seriously affected reproduction of Saguaro and Cardón cactus.

Crested or cristate forms of Saguaro are rather common in the Kino Bay area, but uncommon in Arizona. One or more crests may form at the top of a stem or on the arms of a cactus. Flowers and short stems are sometimes produced on the crest.

In the photo below, a crest has formed on the stump of a broken Saguaro. In the desert around Calle 20 Sur, the author has seen a number of tall Saguaros with broken tops that have forked — one fork forming a normal extension of the trunk and the other forming a crest. In this case, it appears that a broken trunk occasionally causes a crest to form. With the exception noted, the cause of cresting remains unknown, but it does not appear to harm the cactus.

Indians native to the Sonoran Desert consider the Saguaro fruiting time (June and July) to be the beginning of their new year, and celebrate the event. The fruit is gathered with long poles (often Saguaro ribs), the skin removed, and the juicy red pulp and seeds are eaten fresh; or the seeds are dried and stored. The seeds are high in protein and the pulp is high in sugar. Wine is made from the pulp and used in an annual celebration by the Seri around the first of July. Saguaro ribs have been used in daub-and-wattle shelter construction by Indians and Mexicans.

Gila woodpeckers and flickers are responsible for making holes in Saguaros for nests. These pre-built homes are also popular with other birds including small owls, cactus wrens, and even bees. The cactus seals the excavated interior wound area with scar tissue that becomes hard and prevents water loss from the core. When the cactus dies, falls, and decomposes, the scar tissue, called a "cactus boot," is the last to decay. These curious-looking boots in all shapes and sizes are often collected and used for decorations.

Alberto Mónica with crested Saguaro.

Another curiosity is the scar tissue cylinder caused by the Blue Cactus Borer, the larva of the moth *Cactobrosis fernaldialis*. The maggot-like caterpillar burrows into the flesh of saguaros and other cacti, causing small, round scabs on the cactus surface. Behind these scabs (inside the cactus) are small cylinders of callus tissue that are the healed tunnels left by cactus borers.

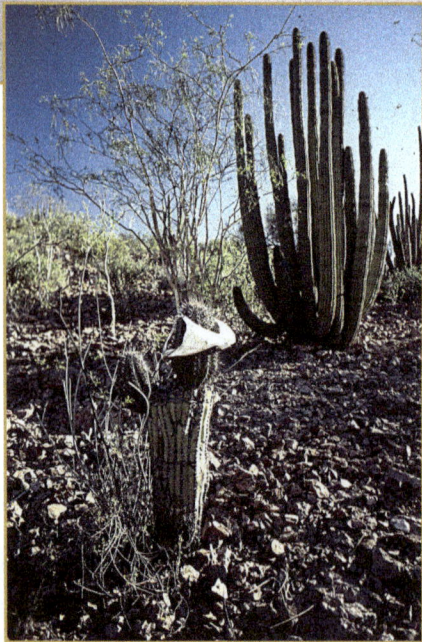

Cactus boot (top) and cactus borer cylinders

Bacterial rot caused by *Erwinia cacticida* turns the flesh of weakened Saguaros into a black, smelly tar-like liquid. One vector for the Erwinia bacterium is the Blue Cactus Borer.

In the photo upper left (taken summer 2008), Saguaros have been transplanted from an area being cleared for shrimp ponds at a shrimp farm southeast of Kino Bay. Nearby, Pinicua trees had also been moved and transplanted. Survival of Saguaros was excellent, but most Pinicuas had died. The practice appears to be a government requirement for protected species.

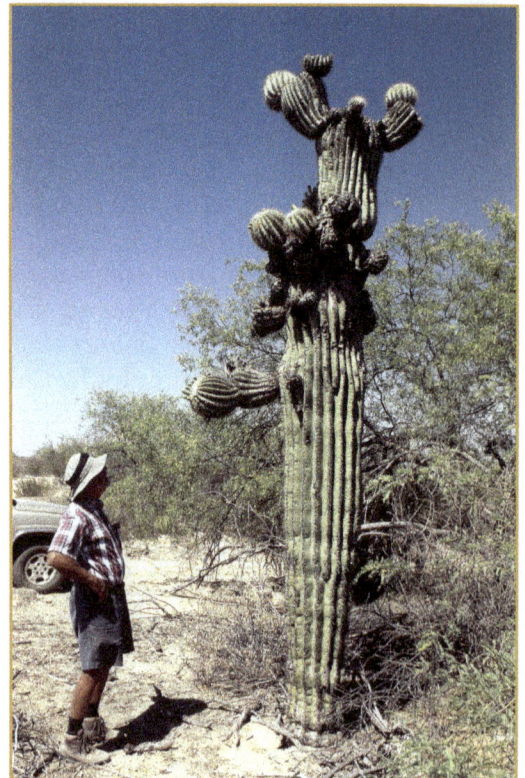

In the photo below right, Dr. Dale Kemmer studies a Saguaro with an odd branching pattern. Even more curious is the double-breasted variety in photo on the left.

WHERE TO SEE IT

Saguaros are common along the highway entering Kino Bay, growing with Cardón cactus.

To see cristate Saguaros, drive to the end of the pavement on Calle 20 Sur. Turn right on the gravel road and travel about ¼ mile, then south to where the road bends east; drive east to a junction, turn south at this junction, and go toward the sea. Many cristate Saguaros can be seen from this road.

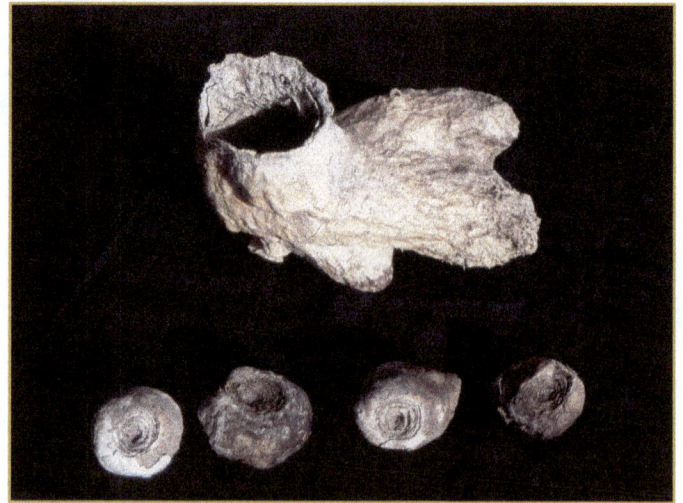

C-10 CARDÓN, Sagueso, Sahueso

Scientific name: *Pachycereus pringlei*
Family: Cactaceae. Cactus family

A large columnar cactus that closely resembles Saguaro. Here on the mainland, Cardón grows on seaside and inland dunes and on well-drained soils within 5 to 10 miles of the sea. Cardón is dependent on maritime weather conditions. On the mainland, it occurs in a narrow band from roughly 50 kilometers south of Guaymas to 40 kilometers north of Puerto Libertád. It also grows over most of the Baja peninsula where it is much more widespread. Annual precipitation for much of its range rarely exceeds eight inches and is frequently less than four inches. A forest of Cardóns is a "cardónal;" thus the name for a fishing village on the coast southeast of Kino Bay.

Large white flowers, similar to Saguaro, appear March through May. Cardóns flower every year, but have more flowers following wet winters. They normally begin to flower two weeks ahead of Saguaros, and fruits ripen by June or July. Most fruits are covered with a brown, felty material, but some are covered with soft, brown spines. The juicy pulp ranges from white to red and contains large black seeds.

Cardóns have the rare characteristic of being trioecious, meaning each plant will have one of three variations of gender: some have only male flowers that produce pollen but no fruit; some have female flowers and produce fruit; and those of the third group have all "perfect" flowers, which means each flower has stamens (male) and pistils (female) and produces a fruit.

Most of the large, multi-stemmed, columnar cacti growing on inland dunes (slight rises) just east of Kino Bay are Cardóns, with a few Saguaros mixed in.

HOW TO TELL THEM APART:

Cardón branches nearer to the ground, and its branches rise at a more acute angle than Saguaro. A mature Cardón has a thicker trunk and many more branches than a mature Saguaro. The base of a mature Cardón resembles a wrinkled leg of an elephant, whereas the base of a mature Saguaro has a rough, tree-bark look. Cardón flowers appear in the top 2 to 6 feet of the stem (photo next page). Saguaro flowers appear primarily as a halo around the top of the stem or sometimes on the upper 1 or 2 feet.

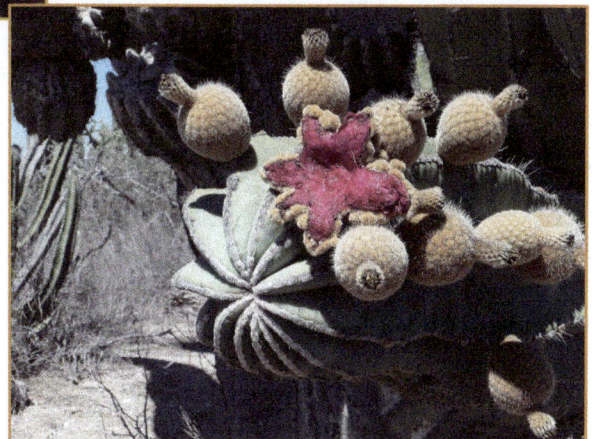

Fuzzy Cardón fruits and a bright red ripened fruit.

65

However, the only positive way to tell them apart is by the areoles. Areoles are tan or gray, slightly raised spots on the cactus rib that subtend the spine cluster. Areoles on Saguaros (photo left) are separate and distinct, with a green area between each areole. Areoles on Cardóns (photo far left) are joined in a continuous vertical line up the ridge of the cactus rib, and there is no green area between areoles.

There are several species of Cardón. This species *(P. pringlei)* grows in Sonora, Baja California, and on the gulf islands, but not in the United States. Saguaros are found in Sonora, Arizona, and southeastern California, but not in Baja California. In contrast to Saguaro, Cardón favors the Baja peninsula where there is a predominance of winter rain as well as the maritime influence of the Pacific Ocean.

Some scientists estimate Cardóns may live two hundred fifty to three hundred years and Saguaros two hundred to two hundred fifty years. Recent research revealed columnar cacti add and retain a vertical growth increment each year, and the age can be determined on height (Turner, et. al. 1995. *Sonoran Desert Plants, An Ecological Atlas*, page 148.).

Cardón is the largest cactus in the world. One on the Baja peninsula measured 72 feet tall. A large Cardón with many arms may weigh 10 tons. Because of their size, Cardóns, Saguaros, and Boojum trees are subject to wind-throw, and many old, tall trees blow down during tropical storms.

Cardóns are attractive as nesting sites for osprey, a fish hawk living near the ocean. In addition to sticks, the birds find a ready supply of nest-building materials along the beach, including fishing line, fish nets, rope, and pieces of clothing. Each year the nesting pair add a little more to the already-workable nest, and in time it may reach a height of 3 or 4 feet. The final product is an engineering feat and a show place of beach trash. However, like the host cactus, these huge nests often meet their fate from high winds that blow away part or the entire nest.

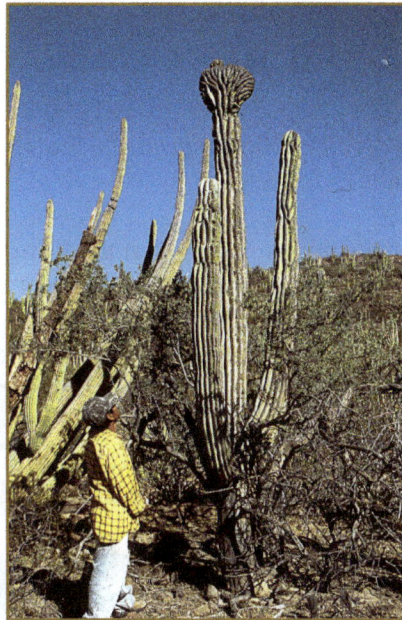

The fruit is high in sugar and was eagerly sought by the Seri Indians who harvested the seeds, buds, and fruit in June and July. The seeds are high in protein and were roasted and eaten, or stored for later use. Juice of Cardón fruit was mixed with charcoal and used in tattooing. The high oil content of the seeds was used to soften deerskins. In the supernatural, the Seri believe Cardón were once people who were their ancestors.

Cresting of Cardón cactus appears to be rare. A crest is an abnormal growth at the tip of the stem (see Saguaro). The Cardón at far left has four small crests and was found near Kino Bay by Corinne Herpel. The one at right was found by the author on a rocky hillside near Tastiota, Sonora, Mexico. They are the only two on the mainland known to the author. Some cristate Cardóns have been found in Baja California.

A related species, Hairbrush Cardón or Little Cardón *(P. pectin-aboriginum),* closely resembles the Cardón of Kino Bay. The distinguishing feature is the fruit which is covered with rather soft, brown, hair-like spines. The fruit, plus spines, is about the size of a baseball. This cactus is common in the southern tip of the Baja peninsula and portions of the southern half of Sonora. In Sonora, Hairbrush Cardón ranges from sea level to the lower elevations of the Sierra Madre Occidental. Several transplants line the entrance to the former Club Med in San Carlos, Sonora.

In the photo at right, cattle grazing was probably the cause of this damage to a small Cardón. Deer and mountain sheep are known to have done similar damage on Tiburón Island. Packrats are suspected where damage is out of reach of ungulates. With maturity, lower spines drop off, making it easy for animals to eat the outer surface. There were many mature Cardóns with similar damage in the area of the picture. Although the injuries happened many years ago, all affected cacti were alive and appeared healthy. (Photo, courtesy of Jim Lyons, Kino Bay.)

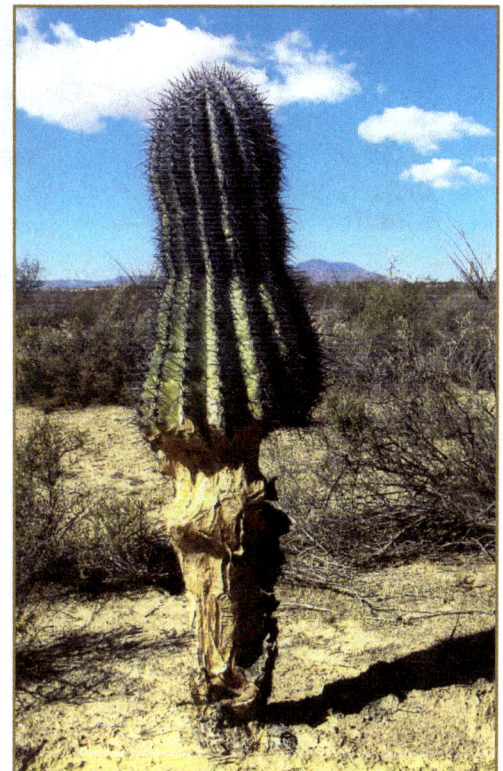

WHERE TO SEE IT
Several Cardóns with osprey nests can be seen at the north boat ramp in New Kino. The most obvious Cardóns are in cardónals (Cardón forests) along both sides of the highway entering Kino Bay.

C-11 SENITA, Sinita, Old Man Cactus, Garambullo

Scientific name: *Pachycereus* (*Lophocereus*) *schottii* var. *schottii*
Family: Cactaceae. Cactus family

A mid-size (8 to 15 feet tall) columnar cactus, often growing in tight groups. Many stems originate from the base and range from 4 to 7 inches in diameter, with five to ten ribs. Mature plants have dense, white, bristle-like spines at the top of the stems, resembling the beard of an old man. No other columnar cactus in the Sonoran Desert has this feature. Senita is often found growing with other columnar cactus, including Saguaro and Organ Pipe Cactus. A grove of Senita is called a "sinitales."

Cactus art

Small, 1 inch long, light pink flowers appear April through August. Flowers open at night to accommodate bat and insect pollinators and close during the morning or afternoon. Marble-sized red fruit with juicy red pulp ripen from July to October. A hawk moth may be the primary pollinator. The moth uses the developing fruit as a food source.

Reproduction may come from seed, branching at the base, or stems touching the ground and taking root.

There are three varieties of Senita cactus in the Sonoran Desert. Ours is variety *schotii,* which has four to seven ribs per stem. It grows in most of Baja California, northwestern Sonora, and barely into south central Arizona. The southern extent in Sonora is just north of Guaymas where it is replaced southward by variety *tenuis,* having six to 13 ribs. Variety *australis* replaces variety *schottii* in the Cape Region of the Baja Peninsula. It differs from the other two by branching well above the base. The other two varieties branch at the base.

Senita colonies are dense and provide shade and a cool place near the ground for rattlesnakes and other wildlife in the heat of the day.

Senita is extremely adaptive. Some grow under the protective canopy of a nurse tree, but most establish and grow well without any cover. The author found one growing in three inches of soil in an abandoned concrete irrigation ditch, and another growing on the prostrate trunk of a live mesquite tree.

Seri Indians ate the fruit fresh. The wood was sometimes used in the walls of brush houses. In the supernatural, the Seri believed Senita was one of the first plants formed. The spirit of vegetation, called Icor, caused the Senita to have a powerful spirit that hovered over the cactus. Calling on this spirit, a person might put a curse on someone or seek help against danger. However, if one called upon this spirit too frequently, the curse might backfire. Good luck was solicited from Senita by wedging clam shells, twigs or other objects into the stems.

A product made from Senita is sold in Sonora as a treatment for sting ray wounds. It is called "garambullo," the Mexican name for the cactus.

Cresting of Senita cactus is thought to be rare. These cristate Senitas were found in a colony a few miles southeast of Kino Bay by Ed and Vy Biskis. (Photos courtesy of Ed and Vy Biskis)

WHERE TO SEE IT
Senita is common in the desert east and south of Kino Bay. Look for the "whiskers" at the top of mature stems.

C-12 SOUR PITAHAYA, Pitahaya Agria

Scientific name: *Stenocereus (Machaerocereus) gummosus*
Family: Cactaceae. Cactus family

A mid-sized (9 to 15 feet tall) upright to sprawling columnar cactus with many stems growing in colonies up to 35 feet across. The stems on mature plants are about 2 to 3 inches in diameter and have eight to nine ribs. The taller stems bend down, lean on each other or touch the ground, and take root to form new plants.

The flowers may be white or white with a purplish tube. They are 4 to 6 inches long and 2 to 3 inches wide. Sour pitahaya flowers throughout the year but most heavily from July through September.

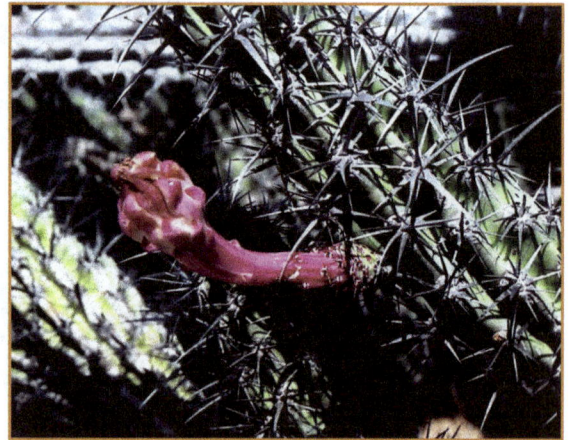

PHOTO LEFT: The fruit is the size of a small orange and has bright red skin and pulp. A small number are produced over a long period. When ripe, the fruit splits open and the spines fall away.

Flowers are night-blooming and close during the day. They are pollinated by moths.

Sour Pitahaya is common over much of the Baja peninsula and grows on San Esteban and Tiburón Islands. A very small population occurs on the mainland just north of Kino Bay near Estero Santa Rosa. Some botanists theorize the Seri Indians may have planted the mainland population as a source of food.

Sour Pitahaya closely resembles Sina Cactus, and both are in the same genus. Here on the mainland, the populations are separated by a distance of 60 kilometers.

The fruit is high in sugar, and the Seri Indians considered it to have the best flavor of the six species of columnar cactus in our region. Wine made from the fruit is said to be stronger than that made from Saguaro, Cardón, or Organ Pipe Cactus. Early Spanish sailors ate the fruit to prevent scurvy.

WHERE TO SEE IT

Sour Pitahaya grows in small colonies on the flat near Estero Santa Rosa, about 2 miles south of Punta Chueca. A few individuals are growing on Cerro Prieto (red hill next to the north boat ramp at Kino Bay), which is the southern extent on the mainland.

C-13 ORGAN PIPE, Pitahaya Dulce

Scientific name: *Stenocereus (Lemaireocereus) thurberi*
Family: Cactaceae. Cactus family

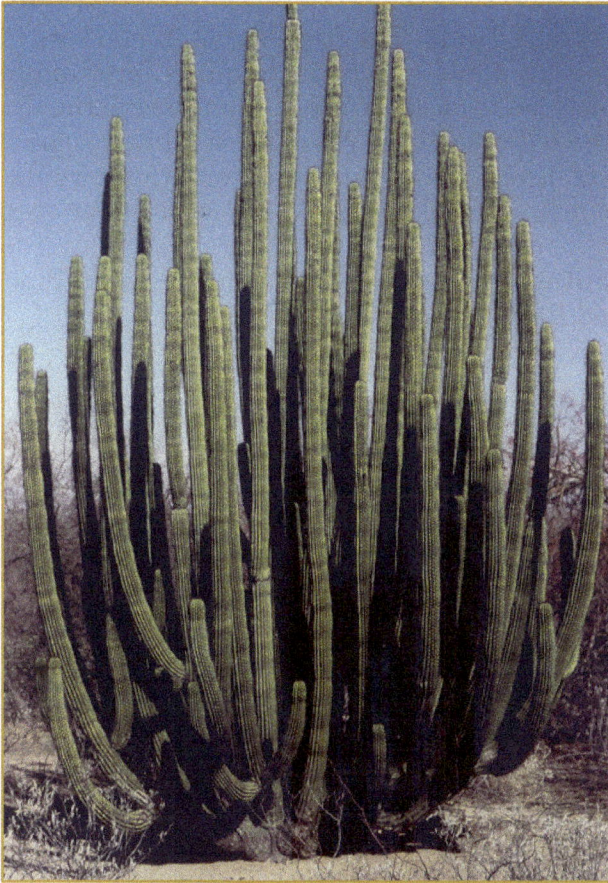

A tall columnar cactus often seen growing with Saguaro. Mature, well-formed plants resemble the graceful pipes of an organ. Most mature Organ Pipes are 9 to 11 feet tall, but some can reach 30 feet. It has a short trunk with five to thirty-five stems. Stems on mature plants are 6 to 8 inches thick, with twelve to nineteen ribs. Stems have a woody core with the ribs connected, unlike Saguaro which has ribs mostly separated.

Organ Pipe grows over much of Sonora and the southern two thirds of Baja California. The only populations in the United States are found in Arizona, where most occur in Organ Pipe National Monument. A few scattered individuals have been found north of there and in a line west of Tucson. The northernmost are a few individuals in the Slate Mountains and at Desert Peak north of Tucson. It is more frost-sensitive than Saguaro.

The flowers are 2 to 3 inches long, funnel-form, and pale pink to lavender, or some are cream-colored and tipped with light purple. Organ Pipe flowers May and June. Flowers open after sunset and close during the day. Bats are believed to be the major pollinators. The flowers have a fruity-skunky odor that attracts bats. Insects are also known

Organ Pipe Cactus flower.

Dried flowers on top of spiny fruit.

71

pollinators. The spine-covered fruit is about the size of a tennis ball, has reddish pulp, and numerous black, shiny seeds. Studies show most plants in Organ Pipe National Monument reach reproductive age when 6 to 7 feet tall.

Organ Pipe prefers the protection of nurse trees to establish and begin growth. Plants have been known to live as long as ninety years.

The name, Pitahaya Dulce, denotes the sweetness of the pulp. The fruit is sold in markets in Sonora and Baja California. A wine is made from the fruit, and "pitahaya daiquiris" are said to be delicious. Indians in Sonora ate the fresh fruit or dried it for later use. Organ Pipe is the only species of columnar cactus where the skin of the fruit is edible. When the fruit was scarce, the gatherer listened for the call of the white-winged dove. Where the doves were, there would be fruit. Torches made of Organ Pipe ribs produced a lot of black smoke and were used for smoking out bees to get honey.

Cresting of Organ Pipe cactus is very rare. The author knows of only two: one in Organ Pipe National Monument and one at the home of Bob and Patty Runner in New Kino.

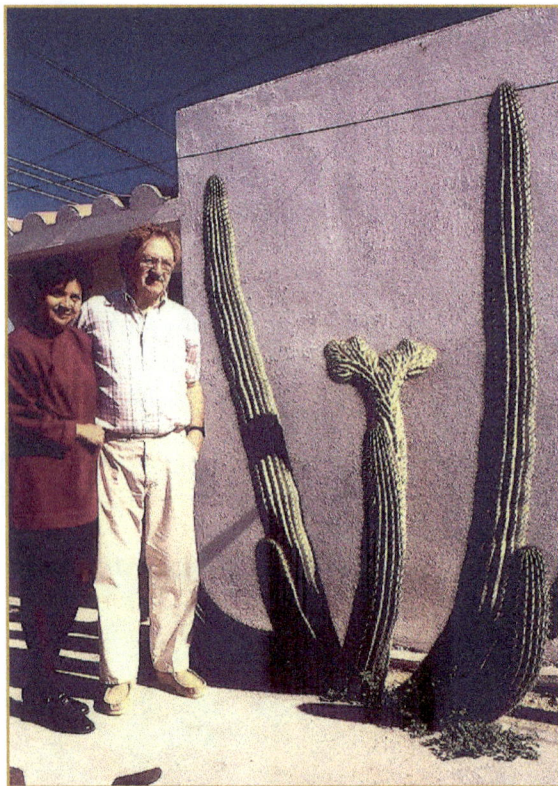

Bob and Patty Runner with their cristate Organ Pipe Cactus at the Kino Bay Realty Office.

Organ Pipe Cactus skeleton. Note fused ribs.

WHERE TO SEE IT
Organ Pipe Cactus is common on the mountains behind Kino Bay.

C-14 ARIZONA BARREL CACTUS, Compass Barrel, Fishhook Barrel, Biznaga

Scientific name: *Ferocactus wislizenii*
Family: Cactaceae. Cactus family

A barrel-shaped cactus usually 3 to 6 feet tall. A few can reach 10 feet and weigh up to 250 pounds. Diameter 16 to 30 inches with twenty to thirty ribs. Stems are unbranched unless injured. The genus name means "fierce" or "wild cactus." Mexicans refer to all tall species of barrel cacti as "Biznaga."

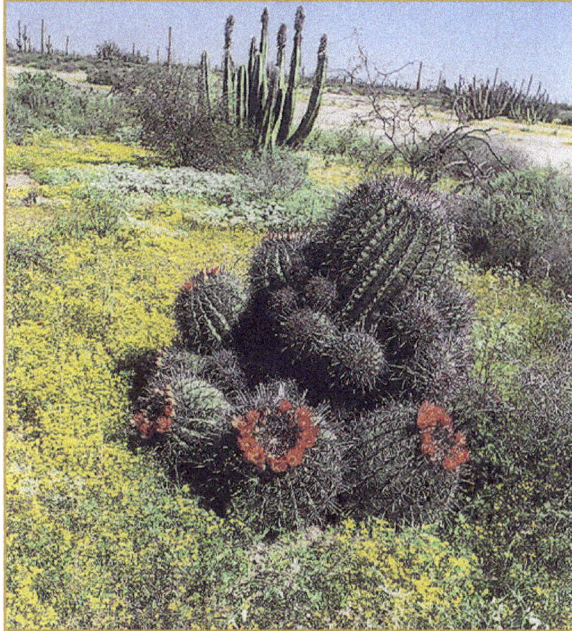

This species is sometimes called "Compass Barrel" because mature ones tend to lean toward the southwest. One theory holds that the greater amount of heat on this side retards growth there, and the larger amount of growth on the opposite side causes the lean. A few plants lean in other directions. Ribs on the south-facing side are 20% closer than those on the north. which allows for cooling.

Three flower colors are possible. Most plants have orange flowers with a strip of darker color on each petal. Ten percent of the plants have either yellow or red flowers. In the experience of the author, each plant produces flowers of a single color. Common flowering time is August through September, with occasional flowering in October. Pollination is mainly by bees. The fruit, called "tuna" by Mexicans, is yellow at maturity, and some may persist on the plant for up to a year.

Shallow roots, 4 to 8 inch deep, extend 10 feet out from the trunk. They are called "rain roots" since they develop after rain and die when the ground dries.

HOW TO TELL THEM APART:

There are four species of barrel cactus in our area. Arizona Barrel is the most common in a 25 mile radius around Kino Bay.

1. Arizona Barrel has four sturdy "central" spines, about twelve "radial" spines (whorl of large spines surrounding the central spines), and twelve to twenty "bristle" spines (whorl of thin, shorter spines beneath the radials). Bristle spines are usually white. Other spines are dark colored, and some may be reddish. One central spine is much longer than the others and is hooked at the end. All other spines in the cluster are straight. The spine "cluster" consists of three spine groups: centrals, radials and bristles. It flowers in late summer.

2. Sonora Barrel *(F. emoryi)* looks much like Arizona Barrel, but lacks bristle spines. It has one central spine surrounded by seven to nine radial spines. The central spine may be hooked or straight. Those around Kino Bay tend to be hooked, while those at Tastiota are straight. The lack of bristle spines is the KEY difference between Sonora and Arizona Barrel. Flowers are yellow, orange, or maroon. Some plants have a decided twist. It grows north and south of Kino Bay, but is less common than Arizona barrel around Kino Bay. The ratio is reversed at Tastiota. It flowers June through September.

3. California Barrel *(F. cylindraceus)* more closely resembles a young Saguaro. The stem is nearly obscured by dense, straw-colored, and yellowish to reddish spines. Flowers are light yellow, occasionally light red. It occurs on Tiburón Island, but its southernmost extent on the mainland is just north of Desemboque. It is the common barrel cactus at Puerto Libertád. It does not grow at Kino Bay. It usually flowers February through April and less often in late summer to early fall.

4. Tiburón barrel *(F. wislizeni* var. *Tiburónensis)* is a variety of Arizona Barrel that is found only on Tiburón Island. It has a central spine that is not hooked and has no radial spines. It flowers in April and May. Arizona Barrel flowers in late summer.

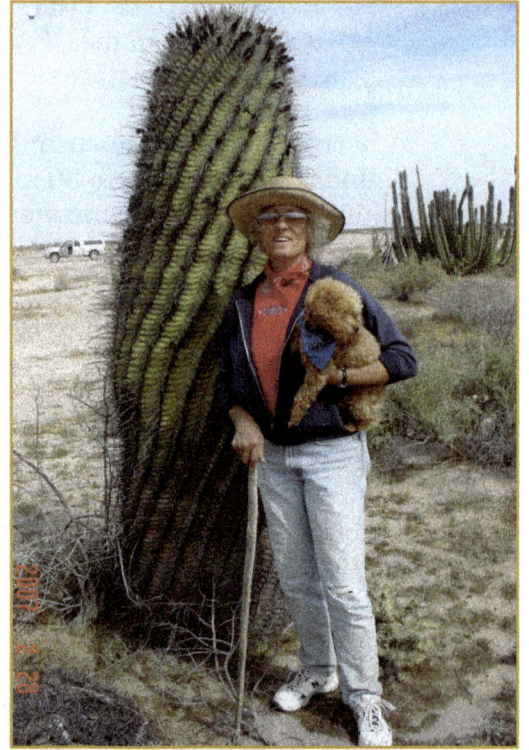

Darlene Lyons and Poopie pose in front of a nearly-eight foot tall Arizona Barrel Cactus found near Calle 20 Sur, southeast of Kino Bay. (Photo courtesy of Jim Lyons, Kino Bay.)

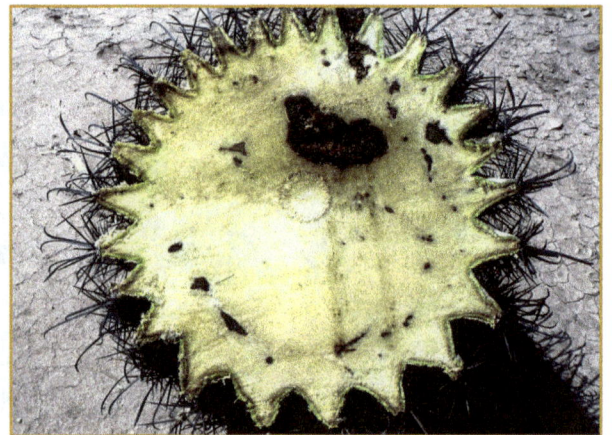

Anatomy of a barrel cactus. Green outer layer is the chlorenchyma which carries on photosynthesis. Yellowish-white inner layer is the parenchyma, or water storage area. Dark spots are damage caused by a large worm.

74

Spine length of Arizona Barrel can vary greatly depending on the age of the plant and probably other conditions. Central spines on most mature plants are about 2 to 3 inches long. Younger plants often have very long spines. The longest central of a young plant found by the author measured 5¾ inches to the top of the hook.

Seri Indians used the spines as awls, needles, and for tattooing. Yellow face paint was made from the flowers. Flowers and buds were cooked in water with sugar and eaten. The fruits were said to be sour like a lemon.

The Seri chewed the water-rich pulp as an emergency water source. It is the only barrel cactus that is reasonably safe to drink; however, eating too much can cause temporary pain in the arms and legs. The pulp of barrel cactus was once harvested and made into candy. The practice was halted in the United States when plant numbers seriously declined.

In the supernatural, the Seri believe a great flood overcame a group of people from the Tastiota region and they turned into barrel cactus. Clouds come from barrel cactus.

The Arizona Barrel on page 73 had 30 branches at the base. It was found near Calle 20 Sur, southeast of Kino Bay. Nurseries induce branching by scoring. Since branching of barrel cactus in the wild is uncommon, this one may have been created by someone with a machete.

The author's daughter, Cathy Maxcy.

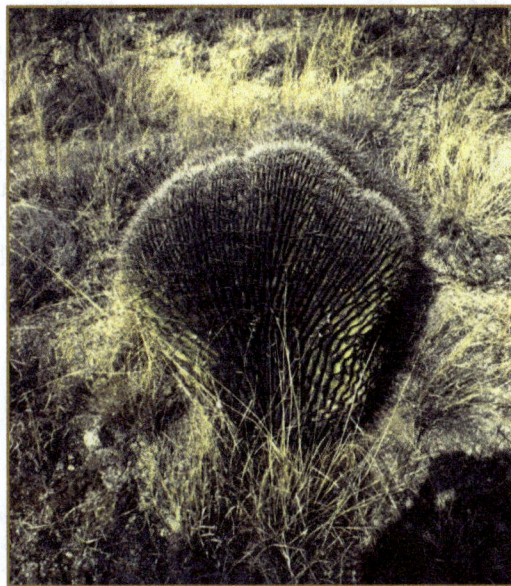

Cristate Arizona Barrel cactus. It had previously flowered and produced fruit. Other barrels in the area were normal. cresting of barrel cactus is uncommon but not rare. Photo north of Benson, Arizona.

WHERE TO SEE IT
Arizona Barrel is rather common in the desert around Kino Bay, growing in association with Saguaro. There are many transplants at homes in Kino Bay. Cristate Arizona Barrels appear to be rare in Sonora.

C-15 SONORA BARREL CACTUS, Coville Barrel Cactus, Biznaga

Scientific name: *Ferocactus emoryi (F. covillei)*
Family: Cactaceae. Cactus family

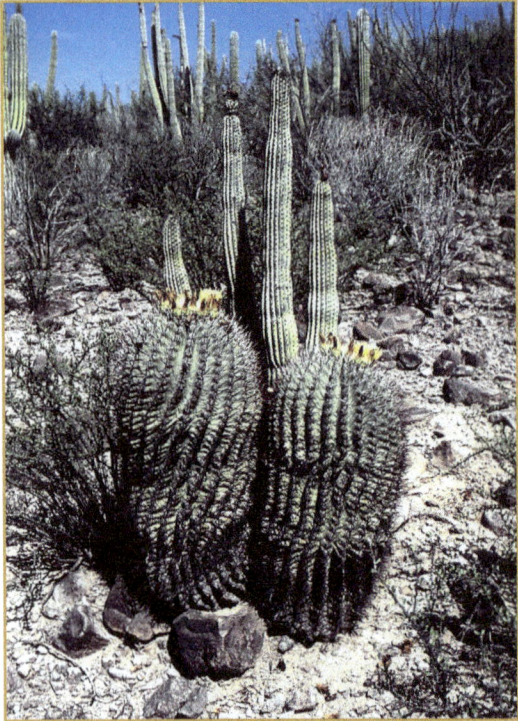

Sonora Barrel is very similar to Arizona Barrel and often grows in the same areas. Arizona Barrel is more common around the Kino Bay area, whereas Sonora Barrel becomes increasingly more common further south. It is the common barrel at Tastiota. Some attain a height of 8 feet.

Flowers are red, maroon, or yellow. The flowering period is June through September.

Sonora Barrel has large hooked or straight central spines, seven to nine thick radial spines, but no bristle spines. Lack of bristle spines makes it easy to distinguish from Arizona barrel.

The Seri Indian name for this species translates to "barrel that kills." Eating the pulp or drinking the juice causes nausea, diarrhea, and temporary paralysis. The Seri used the pulp as a pain relieving poultice. The flowers and fruit are edible. The book, *A Natural History of the Sonoran Desert*, produced by the Arizona-Sonora Desert Museum, says: "No cactus fruit is poisonous, although some are inedible." (Page 205).

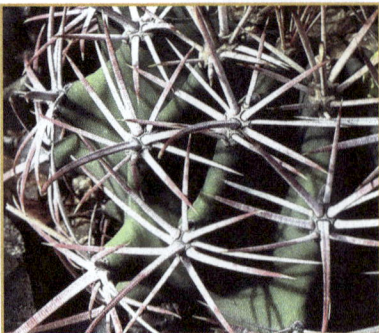

WHERE TO SEE IT:
Sonora Barrel is a common landscape cactus in Kino Bay. There are transplants at Saguaro R.V. in Kino Bay and at Western Horizons R.V. northwest of Kino Bay. There are several along the walk at the Magdalena toll booth that are signed. Both Sonora and Arizona Barrels are common at Tastiota.

C-16 ARIZONA FISHHOOK CACTUS, Pincushion Cactus, Cabeza de Viejo

Scientific name: *Mammillaria grahamii*
Family: Cactateae. Cactus family

A short (1 to 12 inches tall), cylindrical cactus growing solitary or several in a group, often under the protection of a bush. Most are only a few inches tall. The spine cluster has fifteen to thirty radial spines lying flat against the plant and one to three central spines sticking straight out. If there is more than one central spine, the lower one will be stouter and hooked – thus the name. The dense spines nearly hide the stem.

Arizona Fishhook Cactus makes buds during the previous year and then goes dormant. Buds bloom five days after the first two or three summer rains. Flower color is somewhat variable. In our area, flower petals have a purplish or pinkish midrib and whitish margin. The flowers are about 1 inch long and usually borne just below the tip of the stem. Plants may produce a second or third flush of flowers with subsequent rains. Blooming may occur any time between February and July and sometimes as late as October. Very beautiful flowers.

This little cactus is usually found growing under shrubs where they are easily missed by the casual observer. It prefers partial shade, but many larger plants grow in full sunlight.

There are many species of *Mammillaria* in the Sonoran Desert and several different species around Kino Bay. They look much alike and are difficult to identify without the aid of a plant key.

Some plant books show pictures of this cactus with solid pink flowers, which is probably an area difference.

A 16 inch Arizona Fishhook Cactus growing in the open just south of the Kino Bay airport.

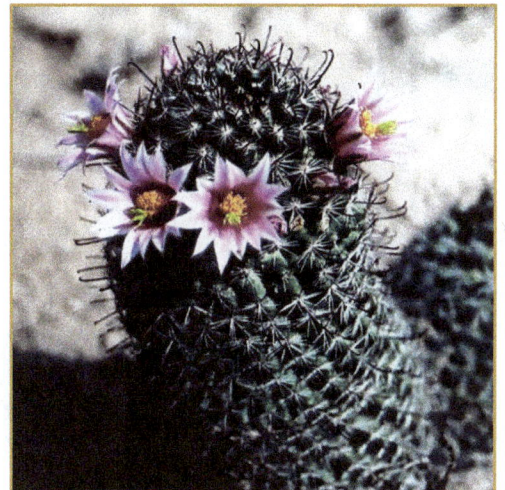

WHERE TO SEE IT
Some very tall (to 16 inches) Arizona Fishhook plants can be seen south of the airport and east of the airport road near Kino Bay. This cactus is fairly common around Kino Bay and at Tastiota. They are most commonly found growing under bushes and are easily seen when in flower.

C-17 CHAINFRUIT CHOLLA, Jumping Cholla

Scientific name: *Cylindropuntia fulgida,* var. *mamillata*
Family: Cactaceae. Cactus family

In our area we have two varieties of Chainfruit Cholla. The variety pictured and discussed here is variety *mamillata* which closely resembles variety *fulgida*. Both occur in the Kino Bay area and occupy some of the same sites. Chief differences are in the stem color and spine abundance. Variety *mamillata* is greener and has fewer spines. Variety *fulgida* has dense spines like Teddy Bear Cholla. Both have distinctive, small, rose-colored flowers and pendant chains of fruit. Variety *mamillata* is the predominant cholla in the Kino Bay area.

In our area, mature plants are usually 5 to 7 feet tall with a few to 9 feet tall. A single, stout trunk supports the scraggly top that seems to have no order of branching. Many branches droop with long chains of fruit. The mature trunk is 4 to 8 inches thick and becomes black and spineless with age.

The flowering period is May to August. The flowers are about 1 inch across with five to eight pinkish or whitish petals that are streaked with lavender. Each plant usually has only a few flowers in bloom at one time.

The fruit is green or yellowish and marble to golf ball size. Older fruit on the chain tend to be smaller than those of the current year. Seeds are few to none. Some plants have viable seeds but they rarely germinate. Reproduction is almost entirely by vegetative means from joints that drop to the ground.

Chainfruit Cholla is easily distinguished from other chollas by its ability to develop fruit on the fruit of the previous season, creating long chains over several years. Chains eventually drop off due to weight or from extended drought. Staghorn Cholla may develop short chains of two to four fruits but usually has only the fruit of the current year. See D-5 in this book.

78

Although widespread in the deserts around Kino Bay, this cactus makes its best growth on flats and lower bajadas with fine textured soils. Dense stands are sometimes found growing on inland dunes where soils are a mixture of sand and silt. Chainfruit Cholla may grow alone or in dense stands.

The Seri Indians ate the fruit year-round. The skin was removed and the fruit roasted or eaten raw. The fruit is high in sugar and rather good. You can eat it fresh, with or without salt. Roasting removes the glochids. Roast over coals for twenty minutes.

Birds, especially white-wing doves, nest in chollas. Packrats cover their stick mounds with cholla joints for protection from predators.

Some species of cacti, including Chainfruit Cholla, develop succulent annual leaves on the new stem growth. These leaves later turn into thorns.

HOW TO TELL CHAINFRUIT CHOLLA FROM TEDDY BEAR CHOLLA:

The picture to the right is of a mature Teddy Bear Cholla. (1) Teddy Bear has branches (joints) that tend to be short and grow upward. Chainfruit has longer branches that mostly tend to sprawl or hang down. (2) Teddy Bear has a single fruit; not long chains as for Chainfruit. (3) Teddy Bear has dense, bright golden spines. Spines on Chainfruit (var. *mamillata)* are sparser and not bright golden colored. Spines on Chainfruit (var. *fulgida*) are dense, but lack the bright golden color and furry, Teddy bear look.

Refer to E-2, Teddy Bear Cholla, In the Puerto Libertád section of the book.

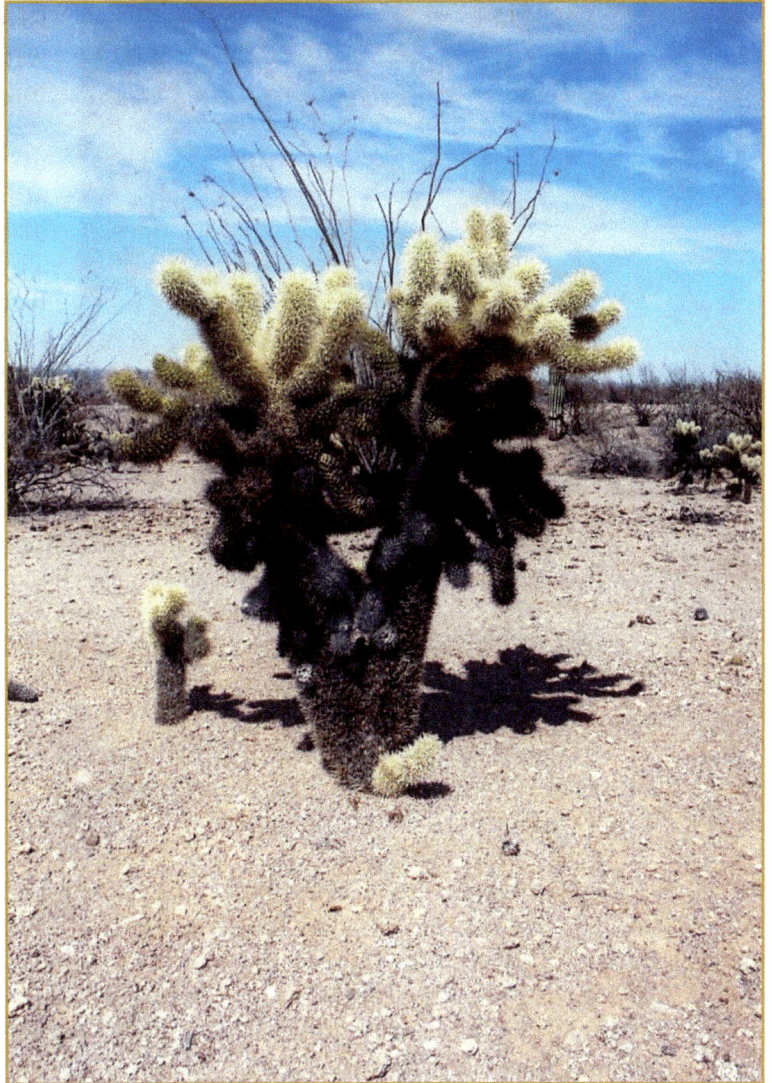

Teddy Bear Cholla near Puerto Libertád. Compare its short stems and upright growth with the longer, cascading stems of Chainfruit Cholla.

WHERE TO SEE IT
Chainfruit Cholla (var. *mamillata*) is common around Kino Bay. Dense stands can be seen growing on inland dunes along the road to Punta Baja, about a mile east of the fishing village. Variety *fulgida* can be seen growing on rocky sites along the road to Punta Chueca.

C-18 DESERT CHRISTMAS, Cholla, Tasajillo

Scientific name: Cylindropuntia *leptocaulis* var. *brittonii*
Family: Cactaceae. Cactus family

Desert Christmas Cholla is the smallest cholla in our area. Most are only 2 to 3 feet tall.
Many grow under the protective cover of a bush. Stems are very thin (¼ to ⅓ inch in diameter), with few spines. Most spines are 1 to 2 inches long, with a few to 3 inches.

Flowers are yellow or greenish-yellow, about an inch in diameter, and crowded along the stems. They open late in the afternoon and close at dark. Bloom occurs from March through June. The flowers are pollinated by flies, insects, and hummingbirds. The stamens are sensitive and retract to the touch.

The mature fruit is finger-like or oval, bright red, about an inch long, and retained on the plant into the following spring. The bright red fruit gives the cactus its name.

Pencil Cholla *(Cylindropuntia arbuscula)* looks much like Desert Christmas Cholla and grows in the same areas.

HOW TO TELL THEM APART:

Pencil Cholla has a few long spines and is a larger plant (5 to 6 feet tall) with thicker (¾ inch) stems. It has small purplish fruit (¾ to 1¼ inch long) versus the bright red fruit of Desert Christmas Cholla.

Desert Christmas Cholla is the most promiscuous of all chollas, crossing with many larger species of cholla to form distinct hybrids. For a list of species it crosses with, see: *Cactaceas de Sonora*, by Aguilar, Van Devender, and Felger. 2000, Arizona-Sonora Desert Museum Press, Tucson, pages 85-86 and plate 51.

Seri Indians removed the glochids from the fruit and ate it fresh.

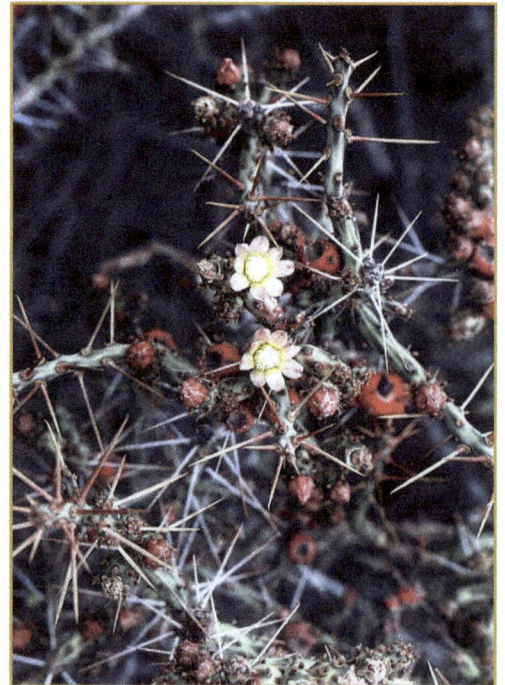

WHERE TO SEE IT

Desert Christmas Cholla has a spotty occurrence in our area. Look for it in Mesquite habitat types southeast of Kino Bay. It becomes more common further south toward (and including) Tastiota. Most grow under a host bush and may not be readily obvious. Look for the bright red fruit that remains on the plant nearly yearlong.

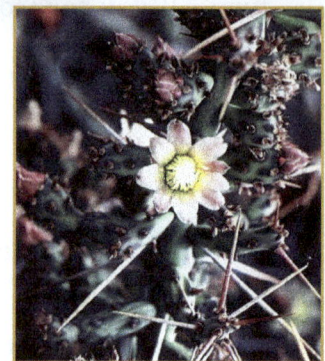

C-19 SONORAN QUEEN OF THE NIGHT, Dahlia-rooted Cereus, Sacamatraca, Saramatraca, Sina de la Costa

Scientific name: *Peniocereus (Neoevensia) striatus*
Family: Cactaceae. Cactus family

A very unusual plant that does not look like a cactus when first sighted. A small, scrambling cactus, 8 to 30 inches tall with no definite form when growing under a host bush. It assumes an upright growth form when growing alone. Fairly common and wide-spread in the area southeast of Kino Bay, but seldom seen. Nearly always growing under a bush where its stems take on the color of the host, making it even less obvious.

Stems are thin, 4 to 8 mm wide, tan or gray, and do not resemble a typical cactus. New stems have short, soft spines, 3 to 4 mm long, that fall off when the stem is older.

The large (3 x 3 inch), trumpet-shaped, white flowers bloom only at night. The flowering period is from late spring to early summer, with fruit ripening in late summer into fall. The fruit is succulent, bright scarlet, round, or oblong. Round fruit is about an inch in diameter. Oblong fruit is 1 to 2 inches long and ¾ to 1 inch wide.

There are shallow striations (ridges and grooves) running the length of the stems, giving the plant its species name, *striata*.

The plant has many large and small, potato and yam-like tubers. Both the fruit and tubers were eaten by Seri Indians. Animals appear to like them too. The author found where an animal (probably a coyote) had recently dug out the tubers and eaten part of one.

Sonoran Queen of the Night is common in Mexico but uncommon in Arizona. It closely resembles and is related to Night-blooming Cereus *(Peniocereus greggii),* a native of southern Arizona, also called Queen of the Night. Each year when flowers of *P. greggii* appear in the deserts around Tucson, the media alerts the public for locations to see and photograph this very beautiful flower. Several local groups host the public for this night event.

The flowers of Sonoran Queen of the Night are even more beautiful than the more popular Arizona species. The flowers open in response to rain, and the spectacular bloom fills the desert night air with an intoxicating scent. After each rain, only a fraction of the flowers on a plant or in a population will bloom on any given night. The flower buds will abort if rain is lacking or insufficient to stimulate flower development. The bright red fruit matures in the fall, which may be the only time a casual observer sees this secretive cactus.

A Sonoran Queen of the Night cactus growing in the absence of a host bush assumes an upright growth form.

Tuberous roots and red fruit.

Flowers of the similar P. greggii.
Photo by John P. Schaefer

WHERE TO SEE IT
The best places to look for this cactus are in Saguaro-Mesquite habitat types with a moderate understory of shrubs such as Wolfberry. This cactus prefers to grow where there is slightly more soil moisture than surrounding sites in the desert. Look for the bright red fruit in late summer-fall.

C-20 WESTERN HONEY MESQUITE,
Torrey Mesquite, Mezquite
Scientific name: *Prosopis glandulosa* var. *torreyana*
Family: Fabaceae (Leguminosae). Pea family
Subfamily: Mimosoideae. Mimosa subfamily

🌿 *Nurse plant. One of the most important plants in the ecosystem of native deserts.*

A small tree, 10 to 30 feet tall, with a broad, rounded, or scraggly crown. Leaves consist of many linear-oblong leaflets, 15 to 22 mm long. Leaves are deciduous in areas where freezing temperatures occur. In the warmer parts of Sonora, old leaves fall and are quickly replaced by new leaves, giving the tree an evergreen appearance. Western Honey Mesquite can live for two hundred years with favorable conditions. Trunks to 30 inches in diameter have been found in the Kino Bay area.

Tiny pale yellow to white flowers grow on a 3 inch long spike. Flowers appear April through August, followed by the beans that eventually dry, split open, and scatter their few seeds. Pollination is mostly from bees. Western Honey Mesquite produces very good honey. Beekeepers move their hives into mesquite groves in the spring.

All species of mesquite have tap and lateral roots that enable them to grow fast and resist drought. Seedlings may remain small for many years while their taproots grow to reach dependable moisture. Mesquite roots penetrating through the roof of Kartchner Caverns in Arizona were estimated to be 80 feet long. At another location, mesquite tap roots were found to be 200 feet deep. Lateral roots extend out from the tree two to three times the height of the tree and have feeder roots near the surface. Mesquites have the deepest roots documented, but 90% of their roots are in the upper 3 feet of soil.

Mesquite roots host nitrogen-fixing bacteria as do the roots of Ironwood and most plants in the pea family. This symbiotic relationship allows mesquite to utilize part of the nitrogen for growth and seed production while building nitrogen-rich organic matter at its base for the benefit of understory plants.

Western Honey Mesquite is one of several mesquite species growing in the Sonoran Desert and the only one in the immediate area of Kino Bay. Velvet Mesquite (*P. velutina*) looks almost exactly like Western Honey Mesquite and shares much of its range in the United States and Sonora.

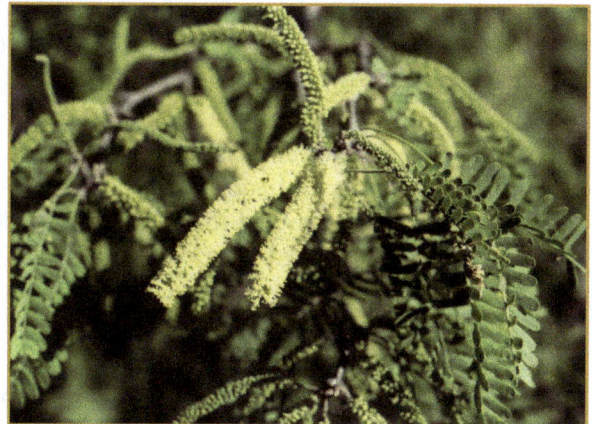

HOW TO TELL THEM APART:

Leaflets and petioles of Velvet Mesquite have a velvet-like covering of hair, and the leaflets are close together. Leaflets and petioles of Western Honey Mesquite have no velvet cover, and the leaflets are spaced wide apart.

A dense stand of mesquite trees is called a "bosque" (pronounced: bose' – kay). Much of the farmland and the fringe of desert along the sea east and southeast of Kino Bay was once mesquite bosques with a wide variety of plants and wildlife, including deer.

Nurse plant: Mesquite is a primary "nurse plant" and plays a major role in the desert ecosystem. The process begins when wind-blown soil builds into a fertile mound at the base of the tree. A variety of other plants gradually establish on the mound, and organic matter and fertility are added from decaying plant material of the host and understory plants. Nitrogen is added from decaying mesquite leaves and beans and from the nodules created by nitrogen-fixing bacteria on the lateral roots of the tree. The enriched soil and protective cover provides a desirable habitat for a variety of plants, including Saguaro. Some species of plants are dependent on this nurse plant relationship for their establishment and survival. These micro-habitats are beneficial for most species of desert wildlife and are essential for the survival of many of them.

When mesquite is removed, many of the dependent plant species (mostly shrubs) at the base of the tree will die, and the mound gradually erodes away. When this habitat is gone, the wildlife component is reduced or eliminated.

Mesquite is an exceptionally useful plant. The wood is hard and attractive. The sapwood is yellow, and the heartwood is a rich reddish-brown. It is used to make furniture, tool handles, posts, and carvings, and for firewood and charcoal. A company in Hermosillo specializes in making furniture from mesquite. The pods and, to a lesser extent, the leaves are eaten by livestock and wildlife. In Sonora, the major use of mesquite is for making charcoal, most of which is exported to the United States for restaurant and home use in barbecues. In an odd twist, once the wood has been converted to charcoal, it looses most of the mesquite smell. Most of the flavor to meat actually comes from smoke from the fat dripping onto the hot charcoal. True mesquite flavor can best be obtained by burning mesquite wood chips or dry pods.

Several mesquite cultivars have been developed for landscape use and are popular in the United States and Mexico.

Mesquite was the single-most important plant for all Indian tribes who lived within its range. It was a primary food source and provided fuel, shelter, material for weapons, tools, fiber for rope and nets, medicine, pestles for grinding, carrying yokes, a black dye for face painting, and many other things.

The beans contain high levels of proteins and carbohydrates. Green pods were picked from the trees and dry pods from the ground. Green pods were mashed and cooked. The most common way to prepare dry pods was to toast them, then grind them on a "metate" or in a mortar. The ground material was placed in a basket and tapped to sift out the flour. The flour was stored or water added to make cakes which were then dried and stored. The flour consisted of the seed minus the hull. An alcoholic beverage was also made from the beans.

WHERE TO SEE IT
Western Honey Mesquite is common along the highway entering Kino Bay.

C-21 IRONWOOD, Palo Fierro

Scientific name: *Olneya tesota*
Family: Fabaceae (Leguminosae). Pea family
Subfamily: Papilionoideae. Papilionoid subfamily

🌿 *Nurse plant. One of the most important plants in the ecosystem of native deserts.*

A tree to 35 feet tall, resembling mesquite. The crown is usually irregular, giving the tree an unkempt appearance. Ironwood distribution is widespread in Sonora, Baja California, southwestern Arizona, and southeastern California. It was once a common tree in all these regions, but their number has been greatly reduced. In Sonora, Ironwood has been reduced by land clearing and harvest for firewood, charcoal, and carvings. In some instances, a few farmers protected them when the desert was cleared for planting crops or range reseeding. The tree is now a protected species in Mexico and Arizona.

The bi-colored, snapdragon-like flowers are pretty but appear dull from a distance. Commonly purple and white, keel petals may vary from pink to violet-purple, with banner petals yellow to white. They appear May through June. The flowers are about ¾ inch long and clustered at the ends of branches. The flowers attract a great many bees. The fruit is a 2 to 3 inch long, twisted pod, with two to four seeds. The pods are eaten by livestock and wildlife.

Branchlets are green and have many short spines, 3 to 11 mm long. Leaflets are oblong to obovate, 5 to 20 mm long, and bluish green.

Ironwood grows slowly and is long lived. Growth rings are often incomplete, so there is no accurate way to age the tree. Estimates of longevity range from three hundred to eight hundred years. The wood is very dense and lignified and will not float.

In Mexico, Ironwood is the preferred firewood as it burns hot and leaves little ash. Mexicans used the wood for tool handles, saddle trees, railroad ties, and mine supports. The Seri Indians began making Ironwood carvings commercially in the 1960s, and original carvings command premium prices today. At present, most carvers are Mexican. Kino Bay is the center for the Ironwood carving industry in Mexico.

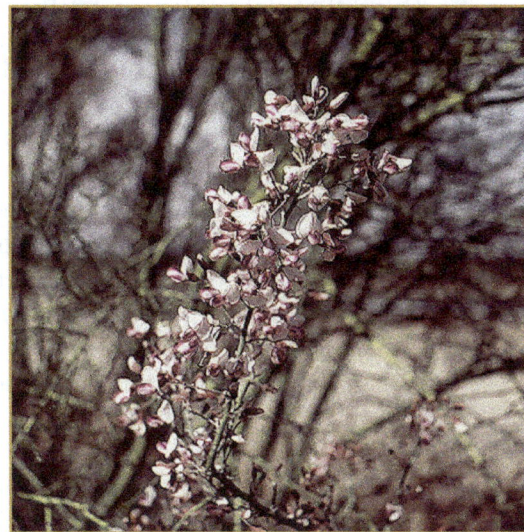

Seri Indians boiled the seeds in two changes of water to get rid of the seed coats and unpleasant smell.

The deer dancer is recognized in both Seri and Yaqui Indian cultures. These carvings are three feet tall. The front carving was made from Ironwood. The other was made from an unknown wood.

The seed coats floated to the surface and were discarded. Cooked seeds were eaten whole or ground and salted. The ground mixture was oily and was said to taste like peanut butter. Ironwood was used for carrying yokes, harpoon points, pestles for grinding, and blades for boat paddles. Bullroarers carved from ironwood were used by shamans and vision seekers to summon the spirits. They are tear-shaped pieces of wood attached to strings and whorled around the head to create a roar. Even arrowheads were occasionally made from Ironwood. William McGee of the Bureau of American Ethnology, previously part of the Smithsonian Institute, noted that in the 1890s the Seri were using Ironwood arrowheads, and therefore erroneously concluded they did not make arrowheads from stone.

Nurse Plant. Studies document 165 plants that benefit from Ironwood as a nurse tree. Like mesquite, Ironwood has nitrogen-fixing bacteria on the roots. It acts much the same as mesquite in its roll as nurse tree. Refer to C-20 (Western Honey Mesquite) for a description of this process.

When not in bloom, Ironwood looks much like mesquite. During the flowering period (May and June) its dusty-purple cast reveals a few Ironwood trees among the mesquite along the highway between Hermosillo and Kino Bay.

HOW TO TELL IRONWOOD FROM MESQUITE WHEN NOT IN FLOWER:
Ironwood leaflets have a bluish-green color versus green leaflets of mesquite. Ironwood thorns are 3 to 11 mm long and fairly numerous along the stem. Mesquite thorns are ¾ to 2 inches long with only a few on a stem. Young mesquites (2 to 6 feet tall), may have thorns 3 inches long or more.

Bill Perrett with a Seri Ironwood pestle found in an excavation along the Playa San Nicholas road.

WHERE TO SEE IT
There is a young Ironwood with multiple trunks along the east-west road opposite the number 4 hole of the Kino Bay Golf Course in New Kino.

C-22 WEDGELEAF LIMBERBUSH, Leatherplant, Sangre de Drago (blood of the dragon), Torote, Matacora

Scientific name: *Jatropha cuneata*
Family: Euphorbiaceae. Spurge family

A thornless bush, 3 to 6 feet tall, with long, thickened branches, curving outward, giving the shrub a fan shape. Stems arise from the base of the plant rather than from a central trunk. The stems are smooth, with a rather attractive gray color. Stems develop numerous spur branches with tiny flowers and clusters of ½ inch wedge-shape leaves. These spur branches are very short and look like little knots along the main stems. Stems are limber and exude large amounts of clear sap when broken – typical of the spurge family. The Spanish name, "Blood of the Dragon," refers to the sap which stains skin and clothes a dark color. Careful!!... The stain is not easily removed!

The tiny, ⅓ inch long, pink to whitish, urn-shaped flowers bloom any time after adequate rain. The normal flowering period is July through September.

The species name "*cuneata*" means wedge-shaped, referring to the shape of the leaf. This is an easy way to distinguish it from Ashy Limberbush, a close cousin, which has much larger and rounded leaves. Both species often grow together and are common in the Kino Bay area. Both are prolific seeders and readily establish in yards and vacant lots in Kino Bay. Leaves drop off during dry periods and regrow following rains. Ashy Limberbush is also featured in this book. See C-23.

Wedgeleaf Limberbush is often found growing on harsh, gravelly sites, such as rocky knobs along the road to Punta Chueca. It grows on gravelly plains and rocky slopes, often in association with species of elephant tree (*Bursera*). The Seri know this shrub as "Torote."

This is the only shrub the Seri Indians use to make their famous baskets. A woman selects plants with long, straight branches, removes the bark, and splits the stem into splints with her

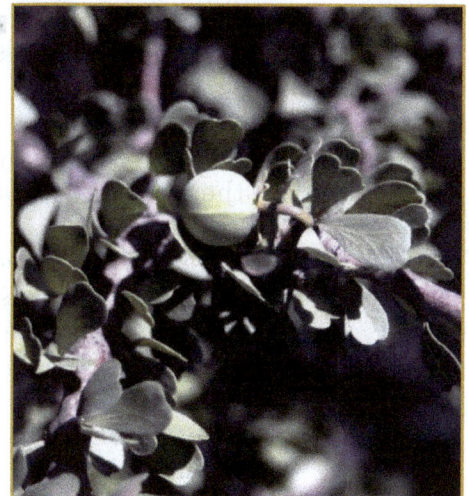

teeth, causing her teeth to have a permanent brown stain. Splints are intricately woven into a continuous coil, and then stitched together. Coils of some are so tight that the basket will hold water. An illustrated description of the process can be found in the book *People of the Desert and Sea: An Ethnobotany of the Seri Indians,* by Richard Stephan Felger and Mary Beck Moser, University of Arizona Press, Tucson.

The baskets are still made in the traditional way, using Wedgeleaf Limberbush and dyes made from native plants and minerals. Stitching is done with a bone awl. Baskets are kept under cover to prevent discoloration from the sun until sold. The white color in these baskets is the natural color of limberbush splints. The red-brown color

Francisca Gastelum Mendez, a Seri weaver at her home in Punta Chueca near Kino Bay. It took about six months to make this large basket with a diameter of 37 inches. In 2008, she was presented with an award for excellence as an Indigenous artisan by the Governor of Sonora.

comes from a dye made by boiling the bark of the lateral roots of White Ratany (see E-6).

Seri baskets are prized by collectors and command premium prices. However, they are very labor-intensive and provide little profit to the gifted artists.

WHERE TO SEE IT
Wedgeleaf limberbush is common in the native desert behind Kunkaak R.V. Park in New Kino and along the road to Punta Chueca, north of Kino Bay.

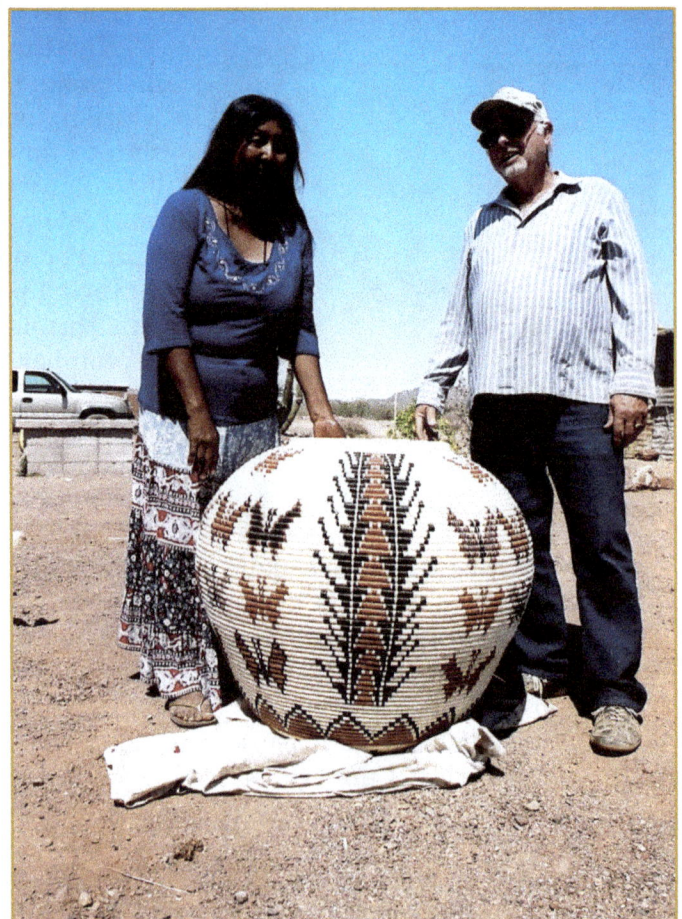

Bob Crowley with Seri weaver. A large basket like this may take a year to make and would sell for several thousand dollars.

C-23 ASHY LIMBERBUSH, Lomboy, Tortillo (little bull)

Scientific name: *Jatropha cinerea*
Family: Euphorbiaceae. Spurge family

A bush or small tree to 20 feet tall. In the Kino Bay area, it is commonly a bush, 4 to 6 feet tall. Many thickened, limber branches arise from the base of the shrub rather than from a central trunk. Branches are long, smooth, thornless, yellowish, grayish or brownish, and rather attractive.

Ashy Limberbush is closely related to Wedgeleaf Limberbush and grows in many of the same habitats. The branches are completely smooth and do not have the small knots along their length as does Wedgeleaf Limberbush. Both are prolific seeders, and seedlings commonly sprout in the yards of homes in Kino Bay.

Small, ⅓ inch long, pinkish tubular flowers may appear any time following rains. The most common flowering period is August to November. The attractive green leaves are rounded, about 2 inches across, and can appear as soon as one day after adequate rain. Leaves turn yellow and drop off during dry periods.

Stems of Ashy Limberbush exude a clear to slightly yellowish sap that can permanently stain clothing. The sap is toxic and was used by the Seri Indians to poison their arrows, although they preferred the poisonous qualities of the sap

from Mexican Jumping Bean (*Sebastiania bilocularis*). The Seri were greatly feared by the Spanish and other Indian tribes for their use of poisoned arrows.

Indians on the Baja peninsula used the sap from Ashy Limberbush to treat chapped or sunburned lips and superficial wounds and to stop bleeding. In some areas where the plant grows tall, the branches are cut and planted to make living fences.

WHERE TO SEE IT
Desert lots behind Kunkaak R.V. Park and vacant lots in Kino Bay townsite. Common around Kino Bay.

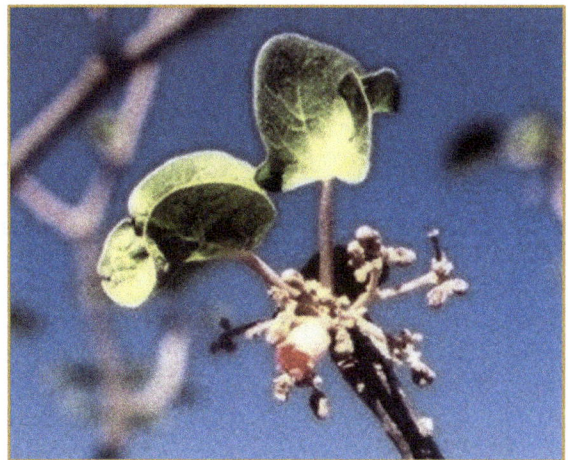

C-24 LITTLE-LEAF ELEPHANT TREE, Torote Colorado (reddish bull)

Scientific name: *Bursera microphylla*
Family: Burseraceae. Torchwood family

A small tree to 20 feet tall, with enlarged trunk and limbs. The trunks are swollen with water storage tissue and covered with sheets of tan bark that exfoliate, especially in spring. The bark of twigs and smaller branches is reddish-brown. The thin bark transmits sunlight to chlorophyll-bearing tissue in the stems, allowing the tree to photosynthesize even when leafless. The sap of the stems and leaves contains turpines which are highly aromatic. This species is widely distributed in the Sonoran Desert and the only common *Bursera* in Arizona. It also grows in southeastern California and most of the Baja peninsula.

The Mexican name, "Torote," is generic and used to identify any tree or bush species having a swollen trunk and/or limbs. The Seri Indians refer to Wedgeleaf Limberbush, the shrub used to make baskets, as "Torote." A stand of elephant trees is called a "Tortal" after the Mexican name for the tree, Torote.

Tiny, inconspicuous, creamy white to pale greenish-yellow flowers appear primarily in July, but may occur at anytime except in the colder months. The pea-size, purple fruit is popular with birds.

There are seven to thirty-one aromatic, shiny, dark green leaflets crowded at the tips of short shoots. The tree may leaf out at any time following the slightest amount of rain, then drop its leaves when the weather turns dry. When a leaf is picked, it releases a harmless spray that smells like turpentine.

HOW TO TELL THEM APART:

There are three species of *Bursera* in the immediate area of Kino Bay. The first two are covered in this book.

(1) Little-leaf Elephant Tree (*B. microphylla*). The trunk and primary branches are noticeably thickened. The leaves have seven to thirty-one narrow leaflets, 6 to 12 mm long. Bark of the trunk and larger limbs is tan or whitish, papery, and pealing in large sheets. This is the most common *Bursera* at Kino Bay.

(2) Red Elephant Tree (*B. hindsiana*). This is the second most common here. The trunk and primary branches are noticeably thickened. Leaves have one, three, or five broad leaflets, in contrast to the narrow leaflets of Little-leaf Elephant Tree. The primary leaflet is oval, about 1 inch across, and much larger than the others. The bark is gray and does not exfoliate.

(3) Torote Prieto *(B. laxiflora)*. The trunk and branches are NOT noticeably thickened. The bark is reddish-brown and not pealing. The leaves are fern-like with five to fifteen leaflets. Uncommon in our area. Occasional on Tiburón Island. It is much more common along Mexico Highway 15 north of Hermosillo and in a large area east of this highway, showing a preference for the area of the Sonoran Desert with a predominance of rain occurring in the summer.

Both Little-leaf and Red Elephant Trees transplant easily. Even mature trees can be transplanted with good success.

There are about one hundred species of *Bursera,* and most grow in Mexico. The aromatic gum of trees in the Torchwood family, called "copal," has long been used in Mexico for medicinal purposes and as incense in religious services. Indians burned copal (not a species from our area) over the heads of Cortés and his men as a sign of friendship when they first arrived. Resin ducts contain these aromatic fluids which are oils of triterpenes and etheria, characteristic of the entire Torchwood family. The family includes Frankincense (*Boswelia* spp.) and Myrrh (*Commiphora* spp.), mentioned in the Bible.

Elephant tree wood is soft and easily worked. The Seri Indians use it to make boat parts, boxes, hand rattles, santos (wooden fetishes), and many other things. The pitch was mixed with animal fat and used for caulking boats and to seal cracks in pottery. The dry wood was considered the best for smoking out bees. The aromatic oils were used in a variety of medicines for the treatment of cuts and stingray wounds. In the supernatural, the tree was considered to have a powerful spirit and was featured in religious practices.

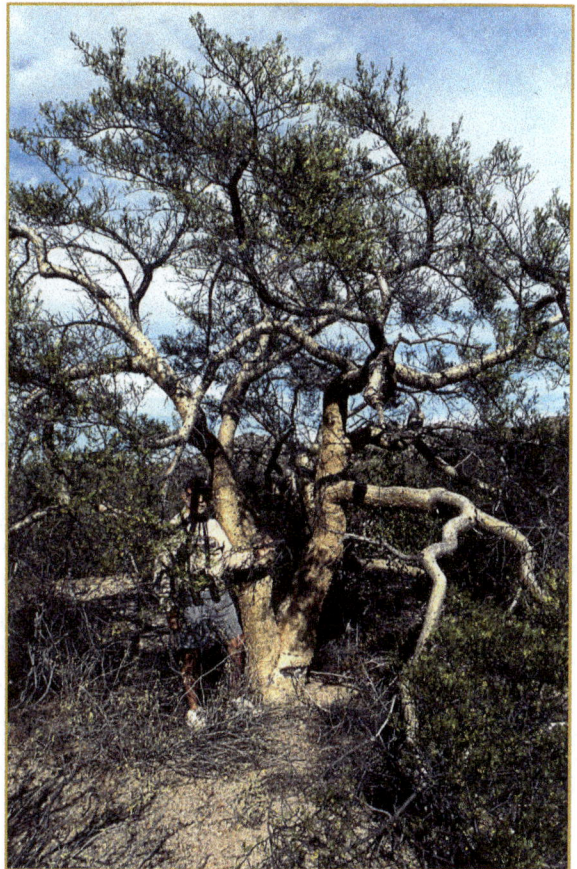

Little Leaf Elephant Tree. Papery, pealing bark distinguishes this tree from Red Elephant Tree.

WHERE TO SEE IT: Little-leaf Elephant Tree is common on the hills behind New Kino.

C-25 RED ELEPHANT TREE, Torote Prieto, Copal
Scientific name: *Bursera hindsiana*
Family: Burseraceae. Torchwood family

A small tree to 24 feet tall with reddish twigs and smooth gray bark on the trunk. The trunk and branches are thickened. Unlike Little-leaf Elephant Tree, the bark does not exfoliate.

The tiny, cream-colored flowers have four petals and are inconspicuous. Flowers appear any time from July to December. Flowering may not occur in some years. The fruit is a reddish, leathery drupe, 5 to 12 mm long, that ripens in the fall and winter.

Leaf shapes are variable. On any tree the leaves may be simple, compound, or both. If compound, there will be one very large, ovate leaflet with two (rarely more) small leaflets at its base. If simple, the leaf will be ovate and up to 1¾ inches long.

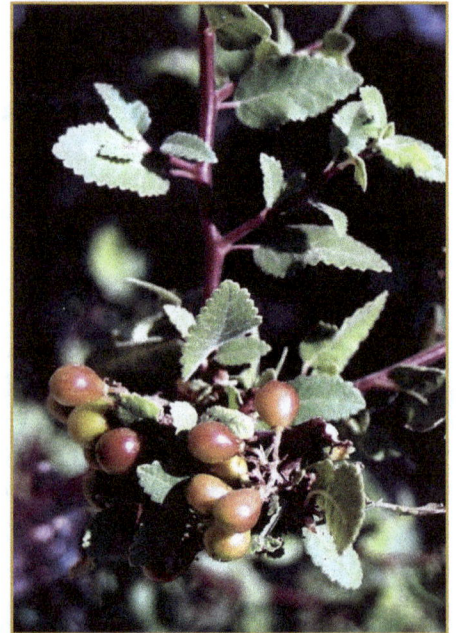

Summer rains stimulate leaf and stem growth. If good moisture continues, the leaves may persist into the fall or the following spring. Leaves are drought deciduous.

Red Elephant Tree grows in washes and arroyos and on gentle slopes and rocky hillsides. It grows in association with Little-leaf Elephant Tree, Adams Tree (Tastiota area), and Ashy and Wedgeleaf Limberbushes.

The Seri Indians used the wood to make small boxes and santos (wooden fetishes). The fruit is eaten by birds.

For a more complete discussion of the *Bursera* genus and Torchwood family, see C-24, Little-leaf Elephant Tree.

WHERE TO SEE IT
Arroyos and rocky hills near the sea just north of Kino Bay. Common along the road to Punta Chueca. When leafless, it can be identified by its smooth gray bark that does not exfoliate like Little-leaf Elephant Tree.

C-26 DOVE PLANT, Broom-wood, Malva Rosa, Malvavisco, Bretonica, Malva de los Cerros

Scientific name: *Melochia tomentosa*
Family: Malvaceae

A perennial half-shrub, 3 to 7 feet tall, with few to many stems arising from the base. Leaves and stems are covered with short, fine hair. A common plant often seen growing by sidewalks and roadsides and in vacant lots around Kino Bay. The lower part of the stem in a "half-shrub" is woody, and the upper part is herbaceous.

The mallow-like flowers are pink to rose-purple, about ½ inch across, and very attractive. They often occur in dense groups of 10 to 15 flowers. Dove Plant can flower year-round following even the smallest rain.

The leaves are ¾ to 1½ inches long, oblong, crinkled, and serrated, with prominent veins. They are blue-green and covered with short hairs. Attractive.

Dove Plant is an opportunistic shrub growing almost anywhere, even in the cracks of sidewalks, so long as moisture is available. When the weather turns dry, the plant drops its leaves and flowers and becomes dormant. If conditions remain dry, the plant dies back to the root crown but responds quickly to even a little rain.

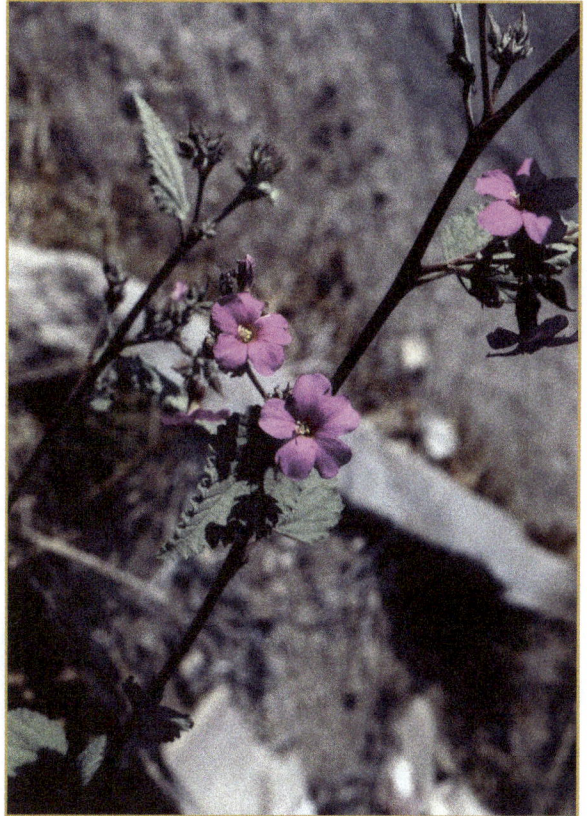

Dove Pant can be transplanted. When watered and trimmed, it flowers almost continuously and makes a very attractive garden plant.

Seri Indians made a reddish-brown dye from the roots. In the Seri supernatural, young girls were cautioned not to look at the flowers or they would become promiscuous.

WHERE TO SEE IT

Most Dove Plants on the mainland grow near the coast. It is an occasional plant along streets and in vacant lots in Kino Bay. It is not obvious when dormant and leafless. Look for it after rain when new leaves and the brilliant flowers appear.

C-27 HUMMINGBIRD BUSH, Chuparosa, Chuparrosa.

Scientific name: *Justicia (Beloperone) californica*
Family: *Acanthaceae.* Acanthus family

Hummingbird Bush is a shrub of desert washes, usually 4 to 6 feet tall, rounded, with dense, whitish stems. It is often leafless, but with a profusion of reddish-orange tubular flowers most of the year, making it one of the most attractive plants of the desert.

The Spanish name "Chuparrosa" refers to the hummingbirds that frequent the flowers. "Chupar" means to suck and "rosa" means rose.

The flowers are 1½ inches long and appear from October through June, depending on rain. The bush is commonly seen with flowers but no leaves. The bright red flowers attract hummingbirds, bees, and butterflies.

The leaves are ovate, ½ to 2 inches long, appearing after rain, but falling soon after conditions turn dry. The plants are usually leafless from October through February.

WHERE TO SEE IT
Dry washes north of Kino Bay and dunes on vacant lots on the north side of Mar de Cortés Boulevard in New Kino.

C-28 BRITTLEBUSH, Incienso, Hierba de Bazo (spleen herb)

Scientific name: *Encelia farinosa*
Family: Asteraceae (Compositae). Sunflower family

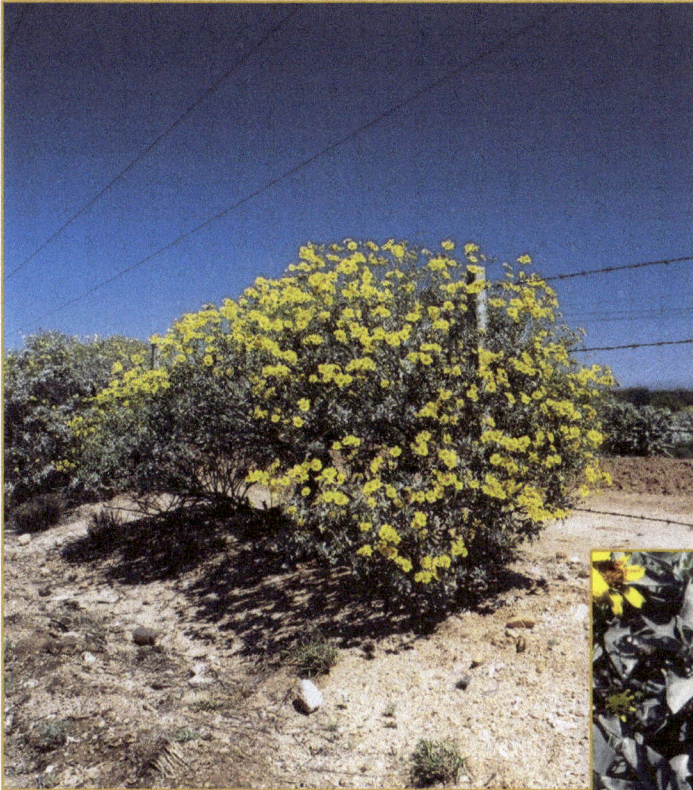

A rounded, densely-branched perennial shrub, usually 2 to 3 feet tall, with daisy-like yellow flowers. A common shrub throughout most of the Sonoran Desert and around Kino Bay. Occasionally used as a landscape plant in Mexico and Arizona. With irrigation, it flowers year-long, and the trunk can reach several inches in diameter. It makes a pretty garden plant.

The Spanish name "Incienso" comes from the golden sap burned as incense in Mexican churches. Break off a dry flower stem and light one end. It has a pleasant odor.

Brittlebush flowers are 1 to 1½ inches wide and consist of bright yellow ray flowers (outer whorl of petals) and disk flowers (flower centers) that are either yellow or dark colored. Note the yellow disk flowers in the photo above and the dark disk flowers in the photo at lower left.

In the past, most plant books recognized two varieties of Brittlebush based on the different colored disk flowers. Botanists now feel this was not a significant characteristic, and varietal names have been dropped.

Flowers appear at the top of a long, slender, leafless stalk. Normal flowering time is late winter to early spring, but Brittlebush may flower anytime after sufficient rain.

Brittlebush may have leaves that are whitish-pubescent (warm season leaves) or green with no pubescence (cool season leaves). The theory here is that leaves tend to have a fairly dense cover of hairs for protection from the sun in warm seasons and the opposite for cool seasons. In the author's experience, this leaf difference does not always relate to the season. Both kinds of leaves may be found in any season.

The leaves of Brittlebush contain a growth inhibitor. Rain dissolves a chemical in the leaves that drips to the ground to inhibit establishment of all but a few annuals. This feature allows Brittlebush to retain a certain degree of site dominance.

Brittlebush often grows in nearly pure stands. It leafs out following periods of heavy rain, with flower production at a later time. Brittlebush is often seen growing with orange-flowered mallows.

During dry periods, the shrub sheds its leaves, becomes dormant, and may be unnoticed except for an occasional one in flower along a roadside. During periods of prolonged drought, Brittlebush may die back to the ground. It regrows quickly following adequate rain.

Brittlebush is grazed by livestock. Heavy grazing gives the shrub a hedged appearance but does not appear to reduce plant abundance.

WHERE TO SEE IT
Brittlebush is common along roadsides and in vacant lots in the Kino Bay townsite. It is easily missed when dormant. Look for it following periods of rather heavy rain that will bring out the leaves and flowers.

C-29 ROCK HIBISCUS, Desert Hibiscus

Scientific name: *Hibiscus denudatus*
Family: Malvaceae. Mallow family

A half-shrub, to 4 feet tall, with beautiful, cup-shaped, light pinkish flowers and thin stems covered with dense brown hairs. Rock Hibiscus is an occasional plant that grows throughout the Sonoran Desert at elevations below two thousand feet. It grows on rocky slopes, flats, and washes, often growing alone or in the protective cover of another bush.

Flowers are light pinkish-white, to 1½ inches wide, with five rounded petals. Some flowers have a dark pink spot at the inside base of each petal. Flowers occur singly or in clusters in leaf axils and at the tips of branches. Common flowering period is January through October, but it can bloom year-round in response to rain. Flower color varies from nearly white in its western range to deep purple-pink in its eastern range. The fruit is a capsule with five chambers that remains on the plant for a time after opening and releasing the seeds.

The leaves are yellowish-green, wooly-haired, oval or elliptical, with toothed margins. The leaves are up to 1¼ inches long, small near the top of the plant, and progressively larger toward the base.

WHERE TO SEE IT

Rock Hibiscus is an occasional plant in vacant lots in Kino Bay town site, along dirt roads, and in desert lots behind Kunkaak R.V. Park It is usually not noticed until in flower.

C-30 COULTER SPHAERALCEA, Coulter Mallow, Annual Globemallow, Mal de Ojo

Scientific name: *Sphaeralcea coulteri*
Family: Malvaceae. Mallow family

An annual forb, usually 2 to 3 feet tall, but can reach 6 feet, depending on soil moisture. Several stems arise from the base. Appears in fall, winter, and spring, following rain. With heavy rainfall, the plants and their bright orange flowers are abundant. Grows on desert flats and slopes. The plant has a wide range of adaptability and is widespread in the Sonoran Desert. It often grows with Brittlebush, where its bright orange flowers make a pleasing contrast with the yellow flowers of Brittlebush. (Photo lower right.)

Mal de Ojo (bad eye or sore eye) is a generic name used by Mexicans to describe several different plants whose pollen can cause sore eyes. Most species of mallow are called Mal de Ojo in Mexico. Packrats collect and store the seeds inside their large stick mounds in the desert.

Flowers of Coulter Sphaeralcea are bright orange to salmon-colored, about ½ inch wide, and clustered at the top of the stems. Leaves are oval, slightly grayish and hairy, about an inch long, and shallowly to deeply three- to five- lobed.

In dry years, Coulter Sphaeralcea is reduced to small patches or individuals several inches high. In years with abundant rainfall, it grows in dense stands, and individuals can reach a height of 6 feet. The bright orange flowers are very attractive.

Coulter Sphaeralcea closely resembles Desert Globemallow (*S. ambigua*), a short-lived perennial growing in the same sites as Coulter Sphaeralcea around Kino Bay. Desert Globemallow has triangular leaves and many stems growing from a slightly woody crown, in contrast to Coulter Sphaeralcea's oval and lobed leaves. Despite what might seem obvious differences in their characteristics, the two are difficult to tell apart.

WHERE TO SEE IT
Coulter Sphaeralcea is often abundant on abandoned farm land, in mesquite habitat types, and along arroyos around Kino Bay following heavy rains. During years with marginal rainfall, the plants are infrequent and not very obvious.

C-31 PURPLE MAT, Purple Nama, Morada (purple)

Scientific name: *Nama demissum*
Family: Hydrophyllaceae. Waterleaf family

A low, annual forb, 3 to 8 inches tall, growing in large mats following winter to spring rain. Purple Mat is an occasional plant in most years, growing singly or in small mats. With favorable rains, it forms large mats of purple or reddish-pink flowers that carpet desert flats and washes. Found in desert habitats of the Sonoran Desert and western United States at least as far north as southern Idaho.

Leaves are thin, linear, sticky, and about 1½ inches long. Flowers are purple to reddish-pink, trumpet-shaped, 3/8 inch wide, and grow on short stalks. The primary flowering time in this part of the Sonoran Desert is February to June; however, an occasional plant can be found following summer monsoons. This little forb stays green and in flower much longer than most desert annuals when conditions turn dry.

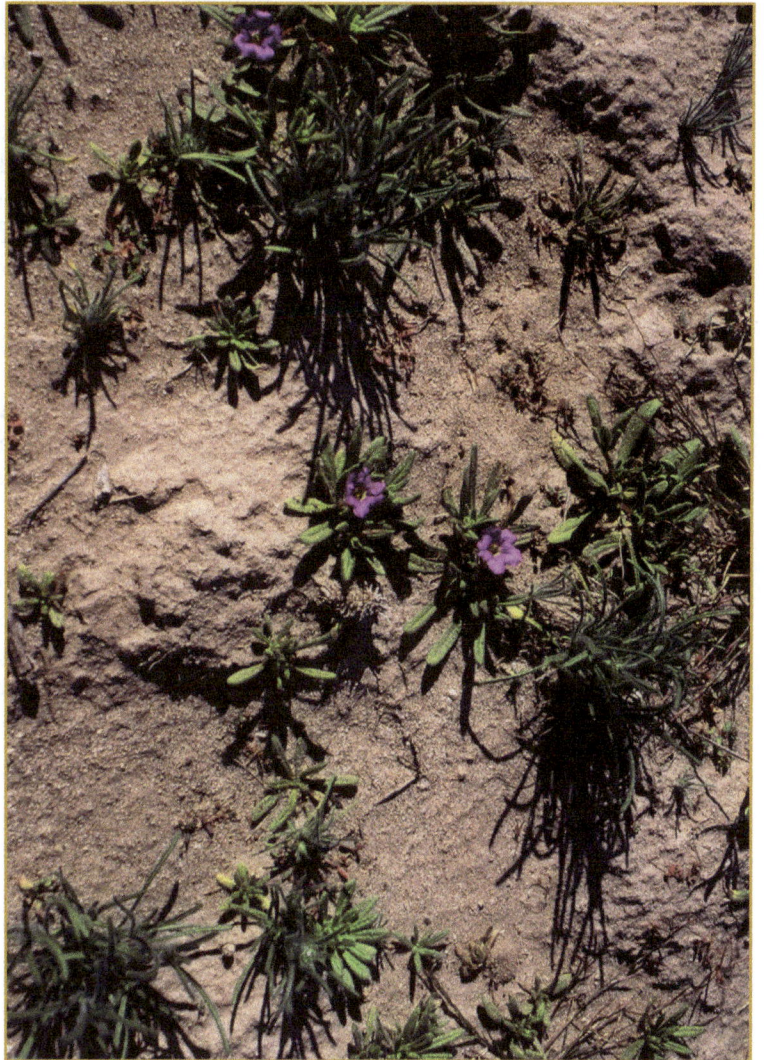

WHERE TO SEE IT
Vacant lots in the Kino Bay townsite and on desert flats east and southeast of Kino Bay. Look for it following winter to spring rains.

C-32 DESERT CHINCHWEED, Fetid Marigold, Mansanilla del Coyote

Scientific name: *Pectis papposa*, var. *papposa*
Family: Asteraceae (Compositae). Sunflower family

An upright or semi-prostrate annual, 4 to 8 inches tall, flowering mid-August through December, depending on rain. It is one of the most conspicuous and widespread hot weather annuals in the Sonoran Desert. It appears after the first heavy rainfall of summer, and successive generations may appear with additional rain into the fall and early winter. It may be occasional or absent in dry years, but abundant following heavy rains, when it forms carpets of yellow flowers. Desert Chinchweed grows on sandy or gravelly plains, on dunes, and in playas. This species occurs in all four of the Great North American Deserts.

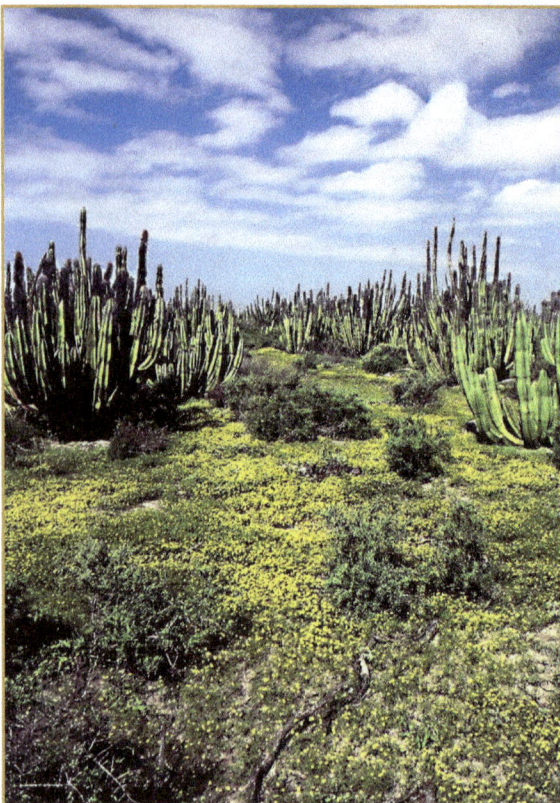

Desert Chinchweed flowers are bright yellow, daisy-like, and about ½ inch across. The leaves are linear and thin, with conspicuous glandular dots on the surface.

Desert Chinchweed has an unpleasant odor when green that becomes stronger as the plant dries. Just walking through a dense patch leaves a strong odor on your shoes and pants – thus the name Fetid Marigold.

In Arizona, Desert Chinchweed and several other plants are co-hosts for the beet leafhopper that carries a virus causing curly top disease in sugar beets.

The author observed Sphinx, or Hawk Moth, caterpillars feeding on Desert Chinchweed plants in August of 2002 near Kino Bay. The caterpillar is about 4 inches long, bright yellow, with black lines running the length of the body. Thousands of caterpillars were crossing Calle 28 Sur and feeding on a large patch of Desert Chinchweed in a playa.

WHERE TO SEE IT
Look for Desert Chinchweed following heavy summer and fall rains. It grows almost anywhere in the desert and on vacant lots in the Kino Bay town site.

C-33 LITTLE-LEAF CORDIA, San Juanito, Vara Prieta

Scientific name: *Cordia parvifolia*
Family: Boraginaceae. Borage family

A shrub, 3 to 9 feet tall, with an open growth and smooth, dark gray bark. Common in the Central Gulf Region which includes Kino Bay. Little-leaf Cordia grows on a variety of sites including desert flats, rocky soils, bajadas, and arroyo terraces.

Large, pretty, white flowers, 1 inch across, appear most anytime following sufficient rain. Peak flowering occurs August to October and March to April. Petals have a distinctive wrinkled appearance. There is nothing attractive about the shrub until it flowers, and then its bright, milk-colored flowers catch the eye and seem out of place in the harsh desert environment. It is the only shrub in the Kino Bay area with large white flowers.

The leaves are ovate to obovate, about ⅓ inch wide and 1 inch long, with veins impressed above and prominent beneath.

Frost-tolerant strains from the Chihuahuan Desert are becoming a popular landscape plant. It is easily propagated from stem cuttings.

There are over 320 species of *Cordia* ranging from Mexico into South America. Some are trees, some are shrubs, and a few are vines. All have the characteristic colorful flower with wrinkled petals that are united part of their length.

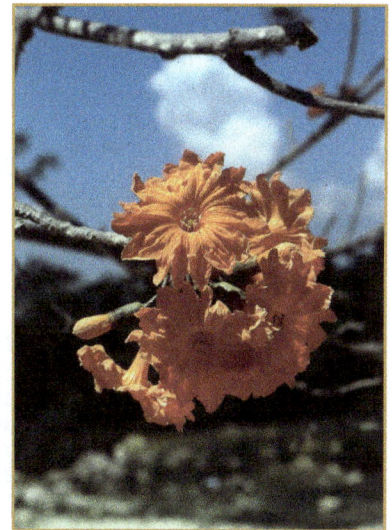

WHERE TO SEE IT
Little-leaf Cordia is easily identified by its bright white flowers. Look for it along the Cat Canyon road north of Kino Bay and on bajadas and arroyo terraces northeast of Kino Bay. It is common on flats east of Calle 4 near Tastiota, growing with Cardón and Palo Adan.

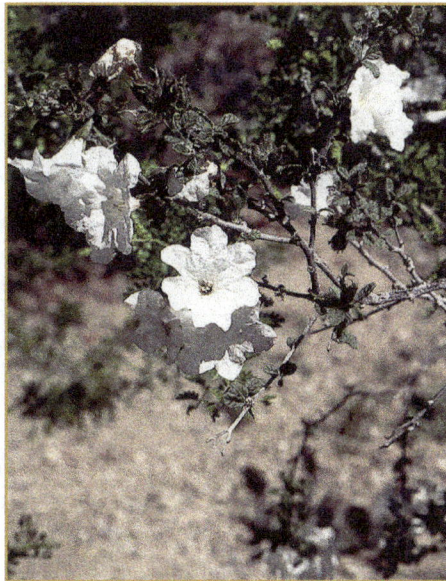

Ciricate (Cordia dodecandra) is a small tree that grows in southern Mexico, the Yucatan peninsula, and Central America. The author photographed these flowers in Belize. The large bright orange flowers appear in winter. Ciricate has beautiful mahogany and yellow-colored wood that is carved to make salad bowls and a variety of other things for the tourist trade.

101

C-34 SWEET ACACIA, Desert Acacia, Huisache, Vinorama

Scientific name: *Vachellia (Acacia) farnesiana*
Family: Fabaceae (Leguminosae). Pea family
Subfamily: Mimosoideae. Mimosoid subfamily

A straggly bush or small tree, to 30 feet tall, with many thorns to 2 inches long. In the dry deserts of western Sonora (our area), it only grows along arroyos and roadsides where there is supplemental moisture. It is more common further east in Sonora where there is greater precipitation. Its eastern extent is the lower elevations on the west side of the Sierra Madre Occidental range.

The leaves are pinnate, medium green, to 4 inches long, with leaflets to ¼ inch long. Flowers are clustered in yellow balls about ⅜

inches in diameter. The flowers are very fragrant and fill the air with perfume when most flowering occurs in the spring. Flowering period is April through November. Foliage and pods are eaten by livestock.

Sweet Acacia originated in the American tropics where the flowers are used for making perfume. It escaped cultivation and now grows wild in many parts of Sonora, Baja California, the southern tier of states in the United States, and south into Argentina.

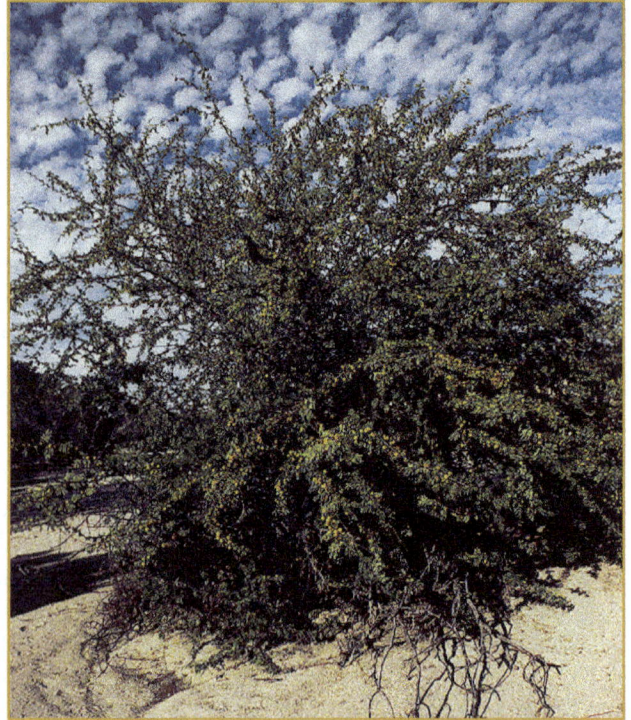

Thornless varieties of Sweet Acacia are popular landscape trees such as this one at a Tucson hotel.

WHERE TO SEE IT
There are a few Sweet Acacia trees growing along the highway between Kino Bay and Calle 36. There are many trees growing with mesquite along the roadside of Calle 36 Norte between 8 and 14 miles north of the junction with the Kino Bay highway.

C-35 DESERT DATURA, Desert Thornapple, Tolache

Scientific name: *Datura discolor*
Family: Solanaceae. Nightshade or potato family

A low, leafy, annual forb, about 16 inches tall, sprawling 1 to 2 feet wide, with large trumpet-shaped white flowers. The flowers are very attractive.

The flower is white with a purplish flush in the throat. In rare instances, flowers may be pale yellow, at least on the ends of the petals. The petals are united into a trumpet-shaped tube 4 to 6 inches long and 2 to 3 inches in diameter. Flowers are nocturnal. They open at dusk and give off a strong, sweet odor. Flowers may remain open on cloudy days. Flowering period is March to October, but the plant flowers throughout the year, depending on rains. The most common flowering time is late summer.

The leaves are 4 to 6 inches long and dark green.

The fruit is a round, pendant pod, about 1 inch in diameter, and covered with long, stout spines. Some people whimsically refer to them as procupine eggs. The green pods turn brown with age, split open, and drop their black seeds.

Sacred Datura, or Jimson Weed (*D. wrightii*), is the only other species of *Datura* in the Sonoran Desert. It closely resembles Desert Datura. Both have beautiful white flowers.

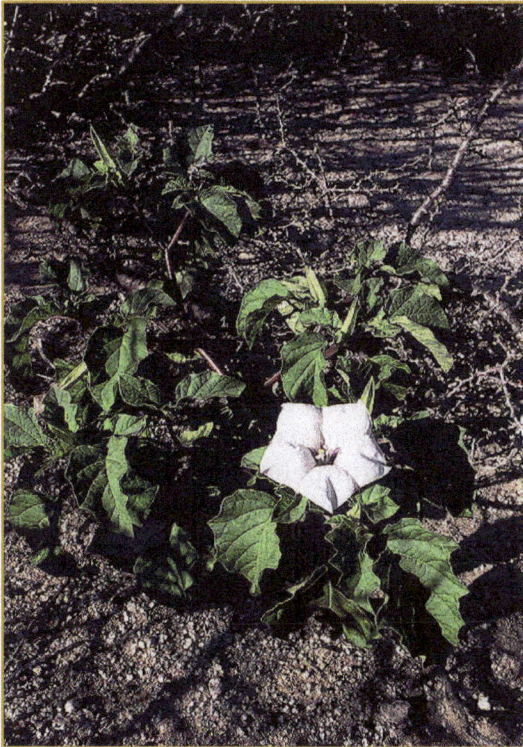

HOW TO TELL THEM APART:

Sacred Datura is a perennial, whereas Desert Datura is an annual. Flowers of Sacred Datura are slightly larger than Desert Datura, and mature plants may be 2 feet tall and 4 to 5 feet across. Sacred Datura pods have slightly smaller spines, and seeds are buff-colored versus black seeds for Desert Datura. Sacred Datura does not grow at Kino Bay but is an occasional plant at higher elevations of the Sonoran Desert.

Indians made a narcotic drink from the leaves that produced a stupefying effect. All parts of the plant contain toxic alkaloids. There are many cases of people eating or drinking concoctions from daturas that resulted in death.

WHERE TO SEE IT
Desert Datura grows in dry waste places and vacant lots in Kino Bay. It is rather common following periods of moderate to heavy rain. A few plants will appear even in a dry period.

103

C-36 TAR TREE, Palo Brea, Mantecosa
Scientific name: *Parkinsonia (Cercidium) praecox*
Family: Fabaceae (Leguminosae). Pea family
Subfamily: Caesalpinioideae. Senna subfamily

A tree to 30 feet tall but usually to 15 feet tall in the Kino Bay area. The tree is unbranched to a height of 3 to 6 feet. The main branches spread nearly horizontally to form a wide-spreading crown. Twigs have one, and occasionally two, small, straight spines at the nodes. The tree is a bright lime-green throughout, including the trunk. This color and the distinctive branching are key features for distinguishing it from other trees in the genus growing in our area (Little-leaf Palo Verde, Blue Palo Verde and Mexican Palo Verde).

Flowers are 16 to 20 mm wide. The petals are golden yellow, 8 to 15 mm long. The banner petal (large upper petal) often has orange dots or flecks. Flowering occurs from March to May. The pods are 8 to 10 mm wide and 2 to 3 inches long, flattened, narrowed toward each end, and not constricted between seeds.

There are five to eight pairs of oblong leaflets, 5 to 10 mm long. The blue-green foliage is also a distinctive feature of this tree.

Tar Tree's name comes from a waxy coating on the bark that is scraped, melted, and used as a gum for gluing leather objects and furniture. Tar Tree is a popular landscape tree in Sonora and southern Arizona.

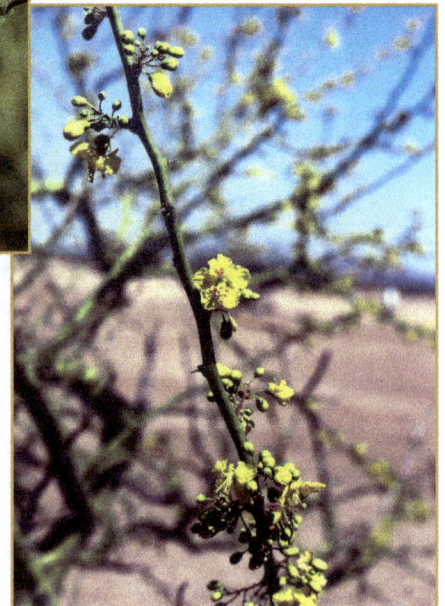

Tar Tree is widely distributed in eastern and southern Sonora and in the southern half of the Baja peninsula. Its western extent on the mainland ends about 20 miles from the coast, and it shows a preference for that part of the Sonoran Desert with a predominance of summer rain.

WHERE TO SEE IT
Tar Tree grows with Western Honey Mesquite and Little-leaf Palo Verde along the highway to Kino Bay between Hermosillo and Siete Cerros (The Seven Hills) west of Hermosillo. There is an occasional Tar Tree along Calle 36 Norte.

C-37 CLEVELAND TOBACCO, Wild Tobacco

Scientific name: *Nicotiana clevelandii*
Family: Solanaceae. Nightshade or potato family

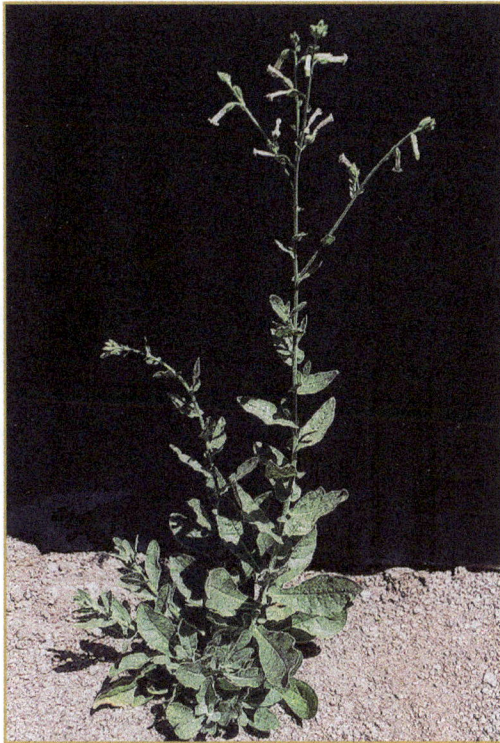

An annual forb, usually 12 to 24 inches tall, with hairy stalks and leaves. The leaves are viscid (sticky). Flowers are white, tinged with violet or pink, narrow, and tubular, about an inch long. The plants have a weedy appearance, although the flowers are rather attractive. The flowering period is February through May.

During most years, Cleveland Tobacco is only an occasional plant around Kino Bay. It is much more common following heavy winter rains, often growing in dense patches along old wood roads in cut-over mesquite sites.

Seri Indians recognized and smoked several species of wild tobacco in their area. The leaves were dried, chopped, and smoked in pipes made from clay or a large tube worm shell.

HOW TO TELL THEM APART:

There are at least three species of wild tobacco at
Kino Bay: The first two were smoked by Seri Indians. It is not known if any Native people smoked Tree Tobacco.

1. Cleveland Tobacco is an annual, has white flowers, and typically grows on the desert floor rather than rocky slopes. It is smaller than Desert Tobacco and has larger leaves. Lower leaves are not "clasping" (base of leaf does not clasp the stem).

2. Desert Tobacco (*N. trigonophylla*) is a biennial or perennial, has white flowers and clasping lower leaves (base of leaf clasps the stem). It is widespread, but typically grows among the rocks on north slopes and along arroyos.

3. Tree Tobacco (*N. glauca*) is a scraggly bush or small, short-lived tree, to 30 feet tall. The stems and leaves are pea-green and hairless. The leaves are large, 2 to 7 inches long. The bright yellow flowers appear year-round. Tree Tobacco is a native of South America that has spread into Mexico and the southern United States during historic times. The leaves resemble domestic tobacco but are highly toxic and have sickened or killed people who ate them as a pot herb.

The leaves of this genus contain the alkaloids nicotine and anabasine, both of which are poisonous to livestock and humans. People eating the leaves of various species of tobacco as "greens" have become seriously ill or died. Nicotine has been used in insecticides.

The genus, *nicotiana*, is named after Jean Nicot, the French ambassador to Portugal. While at court in Lisbon, he was introduced to tobacco which had been brought back from the New World in 1558. He sent seeds of the plant to Queen Mother Catherine de Medicis of France, to whom he owed his position. After returning home to France, Nicot raised tobacco on his country estate and promoted the fashion of smoking among the members of the French court, as Sir Walter Raleigh did later in London.

When Columbus first set foot in the New World, the natives gave him a gift of fruit and "certain dried leaves which give off a distinct fragrance." He ate the fruit, but having no idea of the leaves use, he threw them overboard. A few years later, while visiting Cuba, another explorer named Rodrigo de Jeres saw the natives smoking and brought the custom back to Spain. When his neighbors saw smoke coming from his nose and mouth, they were so alarmed they called in the Inquisition, and Jerez was imprisoned for seven years. By the time he got out of prison, smoking had become fashionable in Spain.

Tree Tobacco

WHERE TO SEE THEM

Cleveland Tobacco grows as an occasional plant on vacant lots or disturbed sites such as land being cleared for construction or along roadsides. Look for it following heavy winter to spring rains.

Desert Tobacco can be found growing on north-facing slopes near Kino Bay and on Tiburón and Alcatraz Islands.

Tree Tobacco grows on disturbed areas where there is more moisture than on surrounding desert sites. It is an occasional plant along roadsides and is common around stock water ponds. It dies very soon after drought conditions dry up its supplemental water source.

C-38 DESERT UNICORN PLANT, Devil's Claw, Gato, Uña de Gato, Torito

Scientific name: Proboscidea althaeafolia
Family: Martyniaceae. Unicorn-plant family

A low-growing, perennial forb with crinkled leaves and bright yellow flowers. The long, hooked, brown pods give the genus its more popular name – Devil's Claw. Flowers are about 1¾ inch across, bright yellow, with brownish-purple specs and dark yellow-orange nectar guides. Flowering can occur practically year-round but primarily May through August. Leaves are dark green, broadly ovate, with shallow lobes.

The pod is a capsule in the shape of a hook that catches the feet of animals to help scatter the seeds. The green, outer cover of the pod is shed, leaving the black inner cover that splits open lengthways to form two hooks and spill the blackish, warty seeds.

Seri Indians peeled the bark of the fleshy root and ate the outer portion (cortex) beneath the bark. The bitter inner pith was discarded.

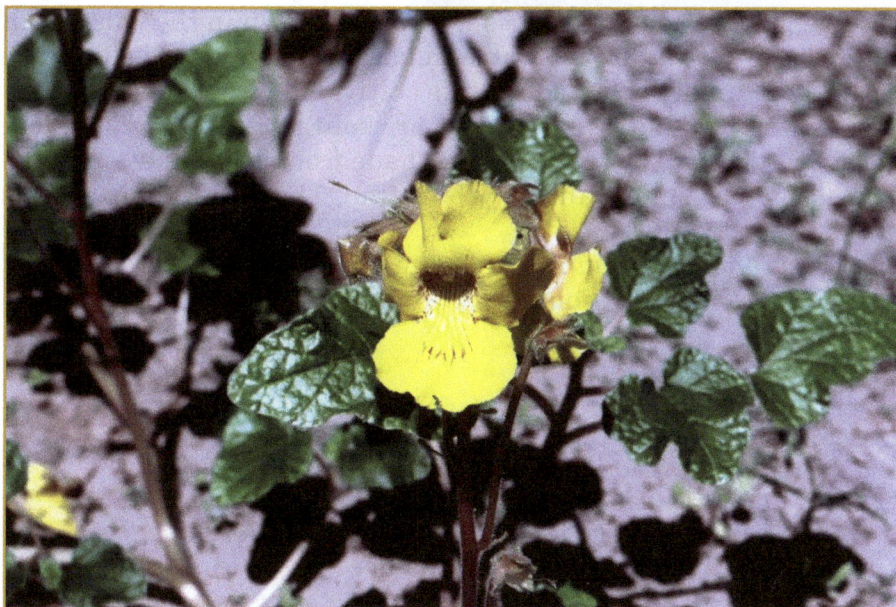

Desert Unicorn Plant begins growth with the start of the summer rainy season and dries up at the close of the season. It is an occasional plant, growing on sandy or gravelly soils and in the bottoms of arroyos. Eye catching when in bloom.

Native people cooked and ate the seeds and immature fruit of a second species, *P. parviflora* (photos, next page). *P. parviflora* flowers are a bright red with yellow center. Seeds were sometimes chewed raw.

P. parviflora, a second species and close relative, also called Devil's Claw, is an annual growing in northern and western Sonora. Its bright red flowers bloom March through October. Pictures taken near Sierra Vista, Arizona.

A third plant, *P. parviflora* var. *hohokamiana*, (not pictured) is a cultivar developed by the Tohono O'odham Indians on their reservation near Tucson. It is not known in the wild and has much longer claws than native plants – a feature desirable in basket weaving. The flowers are pale white with yellow stripes in the throat. The seeds of this cultivated variety are white, versus the black or dark gray seeds of wild plants. The claws are split lengthwise into several splints and used as a black trim or special feature in baskets made from yucca leaves or other light-colored material. These baskets are unique and very beautiful. The Seri Indians have not adopted this feature in their basket weaving.

The domestic variety was named "*hohokamiana*" to indicate its presence in the land occupied historically by the Hohokam Culture of south central Arizona and adjacent Sonora.

WHERE TO SEE IT
The pictures of Desert Unicorn Plants were taken ¼ mile north of the Pemex station in Old Kino. The plants were growing on sandy soil in large openings of scrub brush. Look for it following summer rains. The author has found Desert Unicorn in bloom at Tastiota as late as October, following heavy rains from a tropical depression in September.

C-39 DESERT LAVENDER, Salvia, Bee Sage

Scientific name: *Hyptis albida*
Family: Lamiaceae (Labiatae). Mint family

A shrub to 10 feet tall with many slender stems arising from the base. Dense white hairs covering the twigs, leaves, and part of the flowers give the bush a silvery appearance. The name comes from the tiny lavender-colored flowers.

The flowers are lavender-blue, fragrant, only a few millimeters long, and in dense clusters at the ends of the stems. The primary flowering period is October through May, but the shrub flowers nearly year-long in response to rains. Bees appear to be a primary pollinator.

The leaves are ovate and grayish or whitish to olive green, depending on moisture levels. Leaf blades are mostly ¾ to 1¼ inches long, sometimes to 2 inches. The leaf edges are serrated.

In our area, Desert Lavender is a major component of arroyos. It also grows on rocky slopes in association with the elephant trees and limberbushes. It is widely distributed in the Sonoran Desert.

Desert Lavender flowers and leaves.
Photo courtesy Bert Wilson, used with permission of www.laspilitas.com.

In Seri mythology, Desert Lavender was one of the first plants created. Seri women make colorful little cloth bags filled with Desert Lavender flowers, with a string to hang around the neck. They sell them as good luck charms. The flowers and leaves have a fragrant, mint smell.

WHERE TO SEE IT
Rocky toe slopes and arroyos just northwest of Kino Bay. It is an occasional to somewhat common shrub along the road to Punta Chueca. Obvious by its silvery color.

C-40 THOROUGHWORT

Scientific name: *Eupatorium sagittatum*
Family: Asteraceae (Compositae). Sunflower family

Note: Flowers in these pictures appear light purple. The true blue color requires slow film (OSA 50) and a blue filter.

A densely-branched perennial vine with small stems and small, powder puff blue flowers growing on a host bush. The flowers and leaves are attractive.

Flowers are a soft pretty blue, ⅓ inch long and ½ inch across. Flowers are numerous, appearing at the top of the host plant. Flowering occurs throughout the year depending on moisture.

The dense, green, delicate leaves are thin, wedge-shaped, serrate, and have a saggitate base (base of leaf is forked). The leaf is shaped like an arrowhead.

Thoroughwort grows where there is a heavy cover of shrubs or trees and a greater amount of soil moisture compared to the drier surrounding desert. It is often found growing in the shade of a fairly dense canopy of mesquite or Salt Cedar (*Tamarix ramossisima*) in areas with a high water table at least part of the year. Thoroughwort tolerates a moderately high degree of soil salts.

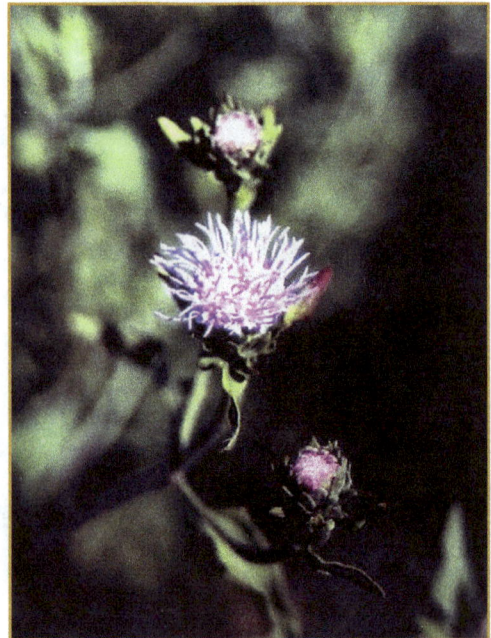

WHERE TO SEE IT
From the Pemex station in Old Kino, drive north on a dirt road about 3 to 4 miles where you enter a dense stand of Salt Cedar, Wolfberry, and other shrubs. This area has a high water table. Look for Thoroughwort vines growing in dense stands of shrubs along the road.

C-41 MENTZELIA, Blazing Star, Stick-leaf, Pega Pega

Scientific name: *Mentzelia adhaerens*
Family: Loasaceae. Stick-leaf or Loasa family

In our area, Mentzelia is a low (8 to 12 inches tall), spreading, annual forb with clinging leaves that have barbed and spinescent hairs. Some plants have trailing branches to 18 inches long.

The stems are whitish, with a papery, pealing surface.

The flowers are bright orange to yellow-orange and about ¾ inch across. Flowers of this species open in the morning. The primary flowering time is October through June, but it may grow and flower any time following sufficient rain.

Leaves are 1 to 2 inches long, dark green, thin, ovate to lanceolate, and shallowly three- to five-lobed or merely toothed.

There are several species of *Mentzelia* in our area that closely resemble each other. Identification can be difficult.

Mentzelia grows on sandy or gravelly soils and on rocky hillsides and in arroyos. Following abundant spring rain, Mentzelia almost carpets the ground on favored sites.

CAUTION! The leaves cling to clothing, leave a green stain, and are very difficult to remove.

WHERE TO SEE IT
Look for Mentzelia after periods of heavy rain. It is common on disturbed sites such as vacant lots and roadsides. The pictures were taken of plants growing on a sand pile at a construction site in Kino Bay.

C-42 ROCK DAISY

Scientific name: *Perityle emoryi*
Family: Asteraceae (compositae). Sunflower family

A pretty annual forb that appears following spring or fall rain. It can be common to abundant if rainfall is heavy. A plant to 16 inches tall with numerous, spreading branches.

The flowers are about ¼ to ⅓ inch across and numerous. Ray flowers are white and disk flowers are yellow. Leaves are ovate to triangular-ovate, 1 to 3 inches long and about as wide, coarsely-toothed to lobed. The lobes are often toothed. The plants can bloom any time, but the main flowering period is November through May.

Rock Daisy is often one of the first annuals to appear after spring rains and the last to die after fall rains. Its distribution is widespread in our area except for dunes and salt flats.

WHERE TO SEE IT
Look for Rock Daisy after moderate to heavy rains in spring or fall. It can be common on vacant lots in the Kino Bay town site.

C-43 FAGONIA

Scientific name: *Fagonia californica* subsp. *longipes*
Family: Zygophyllaceae. Caltrop family

A perennial forb to 16 inches tall, spreading up to 40 inches wide, with numerous thin, squarish stems growing in a thick mat. Fagonia is a half-shrub (stems woody at the base).

The numerous purple flowers are about ½ inch across and very pretty. The fruit is a distinctive pod with five wings. Flowering occurs February through April following sufficient rain.

Leaves consist of three leaflets, the middle leaflet larger than the other two. Leaflets on the lower part of the plant are larger than those on the upper portion, with the central leaflet about 2 mm wide by 10 mm long and lateral leaflets mostly only 1 mm wide and shorter than the central leaflet.

Fagonia has many small (3 mm long), sharp, stipular spines at the base of the leaf petioles.

The stems, leaflets, sepals, pods, and flower pedicels are all glandular.

Fagonia grows on rocky hillsides, gravelly plains, and in arroyos.

WHERE TO SEE IT Look for Fagonia after moderate to heavy winter to spring rains. It grows in the rocky barren strip between Kino Bay town site and the mountains north of town. Also common on the mountain by the south boat ramp at Kino Bay.

C-44 CALIFORNIA CALTROP, Mal de Ojo

Scientific name: *Kallstroemia (Tribulus) californica*
Family: Zygophyllaceae. Caltrop family

An annual forb with sprawling to prostrate stems that spread up to 3 feet. It closely resembles, and is related to, Goathead (*Tribulus terrestris*), and both grow on the same sandy sites in our area around Kino Bay.

The flowers are yellow to yellow-orange about ½ inches across. The petals are 4 to 6 mm long, flaring outward. The fruit is a nutlet about 3 to 4 mm across. The flowering period is May to October, closely matching the March to October period of Goathead.

The leaves are dark green, with three to six pairs of leaflets (sometimes seven pairs). The leaflets are 4 to 10 mm long and 2 to 5 mm broad.

The stems are very narrow and covered with appressed hairs.

California Caltrop grows on sandy and rocky ground, desert flats, hillsides, roadsides, and in vacant lots in the Kino Bay town site.

HOW TO TELL THEM APART:

Goathead nutlets have long, narrow spines that stick in shoes and clothing. California Caltrop nutlets have a "beak," not spines, and do not stick to shoes and clothing. Flowers of Goathead are pale to bright yellow and are open only in the morning. Flowers of California Caltrop are yellow or orange-yellow, each flower opening only part of one day.

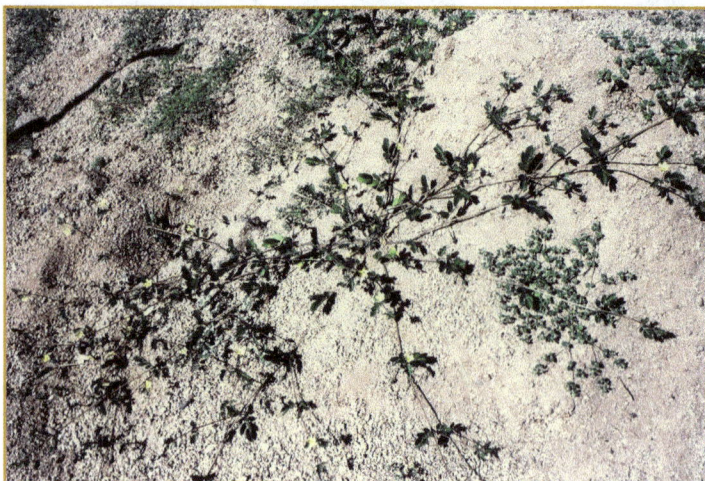

WHERE TO SEE IT
Look for California Caltrop following moderate to heavy summer and fall rains. It grows on sandy sites on roadsides and in the Kino Bay town site. In the author's experience, Goathead grows every year and is rather common, whereas California Caltrop seems to require more rain to germinate and is only an occasional plant. It may be absent when rainfall is scant.

C-45 DESERT BEAN, Frijol

Scientific name: *Phaseolus filiformis*
Family: Fabaceae (Leguminosae). Pea family
Subfamily: Papilionoideae. Papilionoid subfamily

Desert Bean is usually a perennial vine in Arizona, but an annual vine in drier parts to the west, as well as here at Kino Bay. It climbs on host shrubs or trees or sprawls on the ground. A wild bean.

Flowers are pink, about ⅓ inch across. Pods are flat, up to 1½ inches long with brown and black mottled seeds that explode out of the pod. Flowers may appear in any month if moisture is adequate.

Leaves composed of three leaflets, 1 to 2 inches long, thin, highly variable, entire to deeply lobed.

Desert Bean is common and widespread. It grows on flats near sea level, rocky or sandy sites, in arroyos, and on mountain slopes.

Desert Bean is closely related to Teparies and Pinto beans, but not able to outcross and produce fertile hybrid seeds with them under natural conditions. Although very small, Desert Bean is edible and was eaten by the Seri and other Indians when there was an abundance following summer rains. The pods, called ejotes, were also eaten.

Desert Bean growing on a chain link fence in Kino Bay makes a colorful ornamental for a short time. If allowed to go to seed, new plants can be encouraged with minor irrigation.

WHERE TO SEE IT
Look for Desert Bean following moderate to heavy rains. It grows on a wide variety of sites and can be common in the Kino Bay town site.

115

C-46 ALL THORN, Crucifixion Thorn, Crown of Thorns, Corona de Cristo, Abrojo, Junco, Zuccarini

Scientific name: *Koeberlinia spinosa*
Family: Koeberliniaceae (Capparaceae). Crucifixion-thorn or Junco family.

A shrub or short tree, 6 to 9 feet tall (in our area), with numerous, dense branches, and thorn-tipped, yellow-green twigs. All Thorn grows as individuals in our area but can form dense thickets elsewhere.

The flowers are tiny, pale yellow to greenish-white, with four petals (each 4 mm long), four sepals, and eight stamens. The flowers appear in bunches and are quite attractive. The flowers bloom in April, and fruits appear in May. The fruit is a nearly round, green berry, sometimes reddish, about 4 to 5 mm long, and about as wide.

The leaves are awl-shaped (triangular), 1 to 2 mm long, and falling early.

All Thorn has a scattered distribution in the Sonoran Desert. Although it grows on Tiburón Island and around Kino Bay, it is not common here. The southernmost extent in Sonora is near Guaymas. These shrubs grow on sandy or gravelly plains, along arroyos, and on rocky slopes.

HOW TO TELL THEM APART:

All Thorn somewhat resembles Graythorn (*Ziziphus obtusifolia*), a thorny bush common in and around Kino Bay that is also featured in this book (See C-47). Both species have a thick growth and numerous long thorns. All Thorn stands out as a conspicuous shrub or small tree with yellow-green twigs. Graythorn has gray twigs and is a scrambling shrub with no distinct form.

The Seri Indians burned the wood of All Thorn to disinfect their houses and made a tea from the flowers to treat dizziness and intestinal disorders.

WHERE TO SEE IT
Drive north on Calle 36 Norte to the second paved road going west. Turn left and drive west to the end of the pavement. Continue west on a dirt road, across a cattleguard to a sign on the left side of the road that reads "Rancho San Francisco." Turn south toward the rancho. There are many All Thorn shrubs growing here with Ironwood.

C-47 GRAYTHORN, Lotebush, White Crucillo, Bachata, Garrapata, Garambullo, Barbachatas.

Scientific name: *Ziziphus obtusifolia* var. *canescens, (Condalia lycioides), (Condaliopsis lycioides).*
Family: Rhamnaceae. Buckthorn family

Graythorn is a sprawling, thorny shrub with numerous dense, interlocking branches and no definite growth form. It grows as a single shrub or as impenetrable thickets. It can be from 3 to 12 feet high and spreading to 20 feet across. In our area it is usually 3 to 5 feet tall and to 10 feet across. Some large plants grow on vacant lots in the Kino Bay town site where moisture is favorable.

Flowers grow in clusters. Flowers are 2.5 to 3 mm wide, with five sepals, five petals, and five stamens. The petals are white and fall soon after the anthers mature.

The fruit is edible. It is a round drupe, 5 to 10 mm wide, fleshy, purplish-brown, becoming dark purplish-blue when mature. Flowering occurs May through September, but the plants can bloom and set fruit throughout the year in response to rain.

The leaves are alternate on long shoots (new stem growth) and in fascicled groups on short shoots (old stems). The photo below right shows fascicled leaves on short shoots. Leaf shape varies from triangular-ovate to narrowly elliptic. Those in the photo are narrow-elliptic. Leaves are mostly 8 to 22 mm long, 5 to 7 mm wide, with a prominent midrib. Leaves develop and fall quickly, and shrubs are leafless most of the year. For a description of short and long shoot leaves and stems, see C-5 Ocotillo.

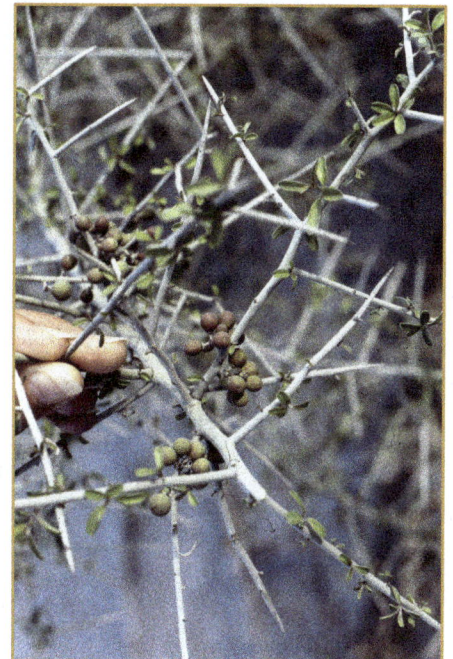

Stems and twigs are gray-green. The thorn-tipped twigs spread at right angles from the stems. Young twigs and young branches are densely hairy but become bare with age.

Graythorn is widely distributed in the Sonoran Desert, growing on flats, slopes, and arroyos.

WHERE TO SEE IT
Vacant lots in New Kino. Also an occasional shrub in the desert southeast of Kino Bay.

C-48 PINK VELVET MALLOW, Malva Blanca, Mariola

Scientific name: *Horsfordia alata*
Family: Malvaceae. Mallow family

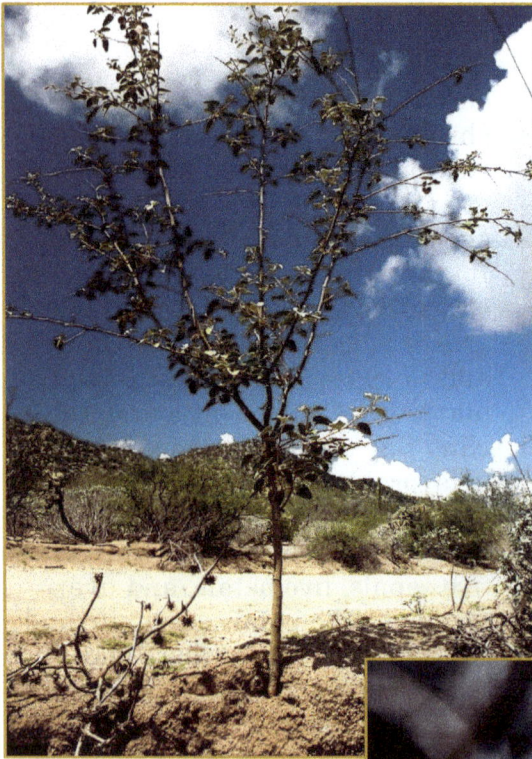

A shrub or short spindly tree, 4 to 18 feet tall (ours mostly 4 to 8 feet tall), with a few thin, upright limbs. Leafing and flowering after rains, but bare much of the year.

Flowers are white to pale pink or lavender, drying to pale lavender or blue, about 1¼ inches across. Flowering period is primarily March to October. The flowers are very attractive.

The larger leaves are mostly ovate with a cordate base, velvety, and 2 to 4 inches long. The uppermost leaves of flowering branches are often lance-shaped, as shown in the photo below.

The stems are slender, tan-colored, thornless, and often covered with dense tan hairs.

Pink Velvet Mallow grows in desert washes and gravely areas. It has a limited distribution in the Kino Bay area but can be found northeast of Kino Bay on gravelly soils.

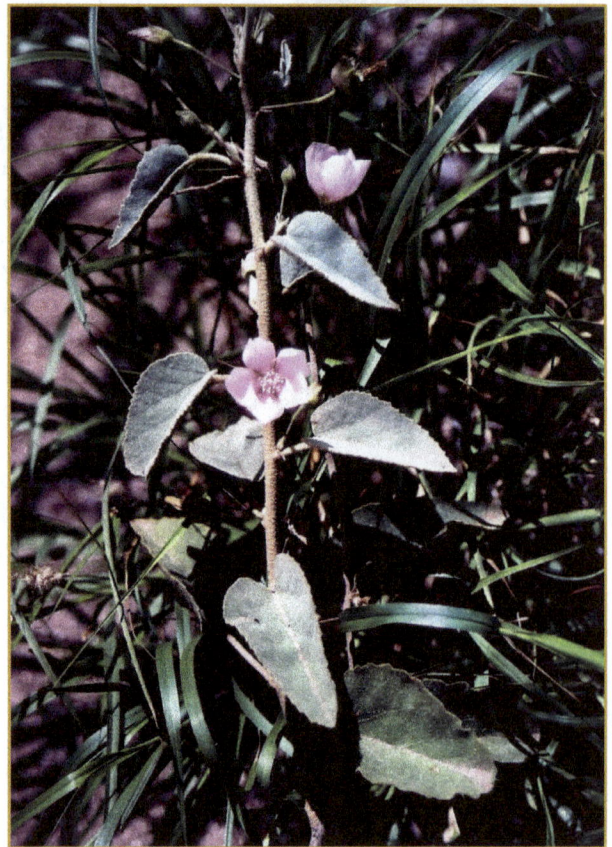

WHERE TO SEE IT
Pink Velvet Mallow is rather common in places along the road between Kino Bay and Punta Chueca, growing on roadside berms. When not in leaf, the plant looks like a dead, spindly little tree. Even with dry winter-spring conditions, you can often find a few reduced flowers on an occasional plant along this road.

C-49 INDIAN MALLOW, Palmer's Abutilon
Scientific name: *Abutilon palmeri*
Family: Malvaceae. Mallow family

A perennial shrub, 3 to 7 feet tall, with a slender, central, woody stalk and sparsely branching thin stems near the top. Indian Mallow is conspicuous from its large, green, velvety leaves and pretty yellow-orange flowers.

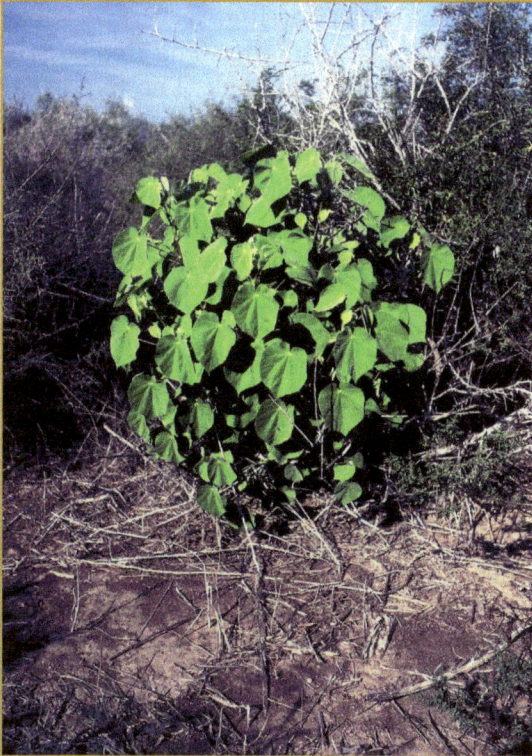

Flowers are bright yellow-orange, about 1 inch across, appearing singly or in small groups at the top of the plant. Indian Mallow can produce leaves, flowers, and fruit any time during warm weather if rainfall is sufficient. In our area, September may be the most common flowering time. After shedding the seeds, the small (½ inch) round pods often remain for almost a year.

The leaves are rounded, deeply cordate at the base, usually pointed at the tip, and covered with a soft mat of velvety hairs. Sizes are variable and can range from 2 to 6 inches long and 1½ to 4 inches wide. Small shrubs may be completely covered with large leaves, such as the shrub at left.

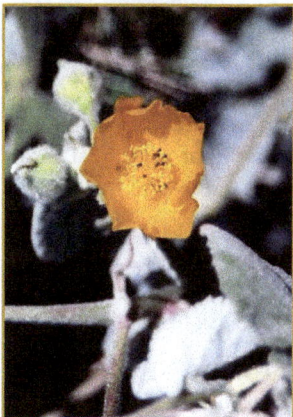

Herbaceous (non-woody) green branches with a dense covering of short hairs are produced from a thin, woody trunk. The branches die with the advent of cold or dry weather.

Indian Mallow prefers to grow in areas with slightly more moisture than the surrounding desert. It grows in arroyos and desert flats, often in association with groups of other shrubs or columnar cactus that provide a degree of shade and a slightly moister micro-climate. Indian Mallow is not common in most desert habitat types but can be rather abundant on those sites described here.

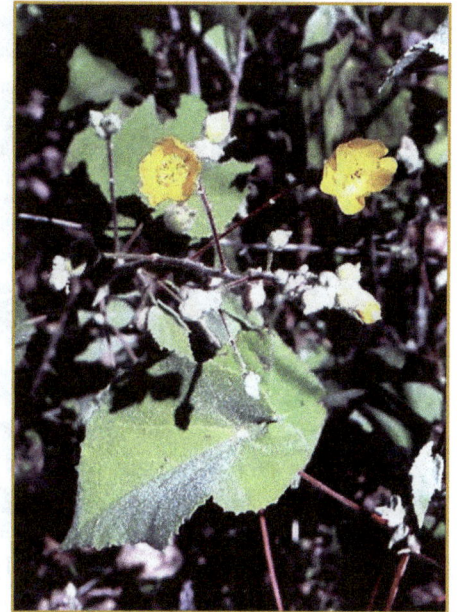

WHERE TO SEE IT
Drive east from Kino Bay on Sonora State Highway 100 to the junction with the San Nicholás Playa road. Park on the north side of the highway and walk north about 50 yards to a group of Cardón Cactus surrounded with many shrubs. There are many Indian Mallows growing here. Some of the taller plants are growing under Cardóns or up through another bush. Look for the yellow-orange flowers on top of the bushes.

119

C-50 PASSION FLOWER, Rosol de la Pasión

Scientific name: *Passiflora arida* var. *arida*
Family: Passifloraceae. Passion flower family

A rounded shrub, or in our area it is more likely a vine climbing on a host tree or shrub, or scrambling on the ground. Passion Flower has many branches from the base, and the stems and leaves are densely woolly. The handsome flowers and leaves look out of place in a desert setting.

The flowers are 1 to 1½ inches in diameter. The five sepals are whitish on the inside and green on the back. The five petals are white, thin, and spreading. The purple-colored features are filaments of the crown. The crown is a structure attached to the base of the sepals and has several rows of purple filaments curving inward in the flower. The filaments give the flower its striking appearance. Flowering occurs February through October.

The fruit (below left) is oblong, about 1 to 1¼ inches in diameter, pubescent, and edible when the surface turns brown.

The leaf blades are about 1¼ inches long and up to 2 inches wide, three-lobed, and wooly. Leaves, stems, and fruits are light green and stand out against the foliage of a host plant.

Passion Flower is not common. It grows in arroyos and on rocky hillsides and flats. In the Kino Bay area, it prefers to grow where there is a good cover of host plants to climb on. It is sometimes found growing on the ground in direct sunlight. All parts of the plant have a foul odor.

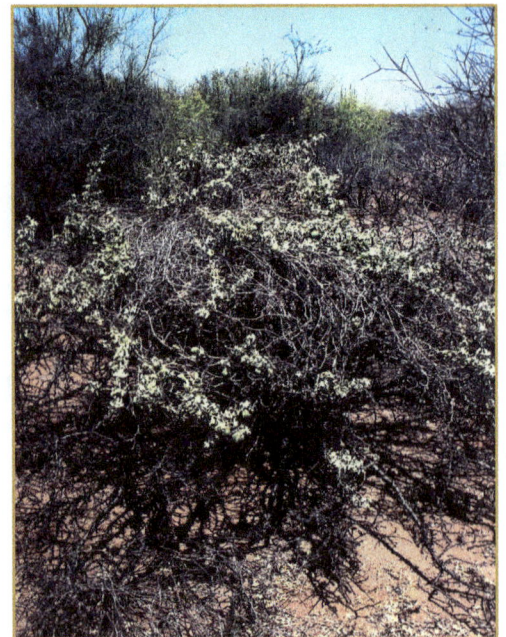

Passion Flower vines on a host shrub.

WHERE TO SEE IT
Flats and gentle slopes between Kino Bay and Punta Chueca with coarse, decomposed granite soils and a moderate to heavy cover of shrubs and trees. An occasional plant can be found growing on the ground in the Kino Bay town site. Despite its delicate appearance, this plant is very hardy and tolerates dry conditions.

C-51 SWEETBUSH, Chuckwalla's Delight

Scientific name: *Bebbia juncea*
Family: Asteraceae (Compositae). Aster or sunflower family

Sweetbush is a short (2 to 5 foot tall), densely-branched, perennial shrub with numerous thin stems and dozens of small yellow flowers on stalks at the top of the plant.

The flower heads consist of disk flowers only. (Compare with Brittlebush which has both disk and ray flowers). Heads are about ½ inch long and ⅜ inch wide, bright yellowish-orange, attractive and fragrant. Flowering heads are on long stalks. Although each plant produces many flowers, most have only a few flowers open at a time. Flowering occurs any time of year depending on moisture. When flowers mature, the plant is covered with dozens of tiny white puff balls of seed like miniature dandelions.

The leaves are linear to linear-oblanceolate, sparse, and often semi-succulent during times of favorable moisture. The leaves are 1 to 2 inches long with size reduced upward in the plant. The leaves are quickly drought deciduous and the plant is leafless most of the year. Flowers and leaves quickly reappear after any good rain.

Stems are dense, slender, brittle, pale green, and without thorns. When the leaves drop, the stems look like reeds. Stems have the ability to produce chlorophyll which allows the plant to continue photosynthesis after the leaves drop off.

Sweetbush grows in a wide variety of habitat types from sandy arroyo bottoms to dry mesas and flats. It is common on roadsides and vacant lots in the Kino Bay town site. It is grazed by livestock and wildlife, including chuckwallas.

WHERE TO SEE IT
Sweetbush is common on vacant lots and roadsides in the Kino Bay town site. It is a common plant of arroyos. Look for the flowers any time of year following rain.

C-52 CHAPACOLOR, Amole, Tinta

Scientific name: *Stegnosperma halimifolium (S. watsonii)*
Family: Phytolaccaceae (Stegnospermataceae). Pokeweed family)

A densly-branched, rounded shrub to 16 feet tall (6 to 7 feet in our area), with arching branches that criss-cross and droop. Some branches droop to the ground. Shrubs growing alone, outside the influence of other shrubs or trees, often have leaves so dense they nearly hide the trunk and primary stems. A pretty shrub that looks somewhat out of place in its dry desert setting.

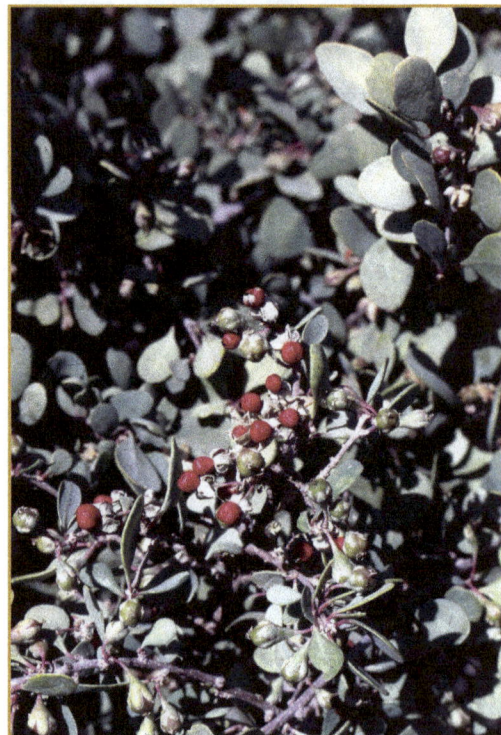

The tiny flowers are about ⅓ inch wide. The petals, stamens, and stigma are bright white. The sepals are greenish and often have a tinge of red. Flowers are showy, dense, and fragrant, attracting many bees. Flowering period is October through May, but Chapacolor may flower any time following adequate rain.

The fruit is a round capsule, 6 to 7 mm wide, green at first and then turning red. The bright white flowers and bright red berries are very attractive.

The leaves are pea-green, elliptic to obovate, rather thick, and ¾ to 2 inches long. Leaf petioles and leaf veins are often red. Leaves are evergreen except in extended dry periods.

Chapacolor is an occasional to locally common shrub along both coasts of the Gulf and in desert washes. It is found only in Sonora, Sinaloa, Baja California, and several Gulf islands including Tiburón. The distribution shows it favors a maritime climate and sites with higher levels of soil moisture than the surrounding desert.

The Spanish word "tinta" means tint or color. The juice of the fruit leaves a brown stain on skin and clothing – thus the Spanish name for this shrub.

WHERE TO SEE IT
There are three Chapacolors within sight of the number 2 hole of the Kino Bay Golf Course. The nearest is 50 feet south of the hole. A large one is growing in the fence on the west side of Calle Cadíz opposite the Prescott College facility in New Kino.

C-53 PALO COLORADO, Granadita
Scientific name: *Colubrina viridis (C. glabra)*
Family: Rhamnaceae. Buckthorn family

A large shrub, to 8 feet tall, densely-branched, with stout twigs branching at almost right angles to the main stems. On large shrubs, the ends of the branches tend to curve outward, giving the shrub a distinct outline.

The tiny yellowish-white, star-shaped flowers are 5 to 6 mm broad. Flowering occurs February through May and September through October. The fruit is a round red capsule, 4 to 6 mm in diameter.

Palo Colorado leaves are obovate to suborbicular, 5 to 30 mm long, 5 to 16 mm wide, thin, and bright green. The bright green leaves set the shrub apart from other species on a site. Leaves fall quickly with dry conditions and reappear following rains. Palo Colorado is leafless much of the year.

The branches are stiff and the wood is very hard. Stems are thornless, but the ends of the twigs are thorn-tipped. The bark is gray. Branchlets grow at nearly right angles to the branches.

The Seri Indians used the stem as a pry bar to dig up Agaves and for fire wood. A tea was made from the roots as a remedy for measles. In Seri mythology, giant girls gathered the seeds. One who was in love strung the seeds into a necklace and gave it to her lover.

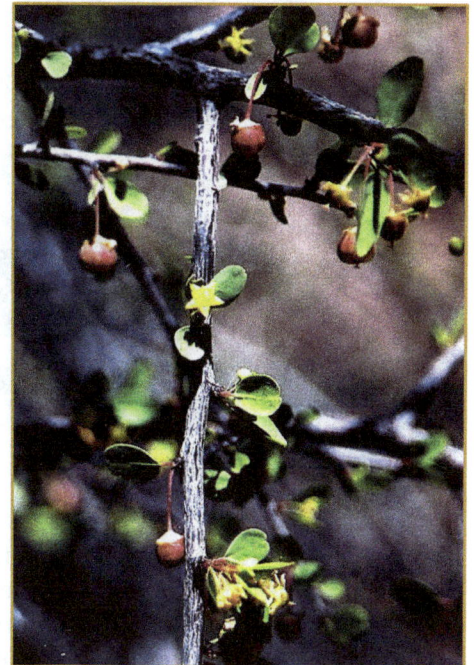

Palo Colorado is one of the most common shrubs on bajadas, plains, and hillsides, and in arroyos northwest of Kino Bay. It can be distinguished by the right-angle branching of twigs, curved ends of branches, and bright green leaves. Palo Colorado grows in association with a wide variety of shrubs and trees. In and adjacent to arroyos, it is found growing with Little-leaf Elephant Tree, Wedgeleaf Limberbush, Ashy Limberbush, Cat Claw Acacia, Wolfberry, Trixis, Balloon Vine, Desert Lavender, Organ Pipe, and Cardón Cactus, to name just a few.

WHERE TO SEE IT
Palo Colorado is common in mixed shrub and tree habitats on bajadas and along arroyos northwest of Kino Bay.

C-54 BEETLE SPURGE, Desert Poinsettia

Scientific name: *Euphorbia eriantha*
Family: Euphorbiaceae. Spurge family

An erect annual or biennial forb, 1 to 2 feet tall, with a single stem at the base and several branches. The foliage is pea-green.

The small male and female flowers are in a cup-shaped cyathium at the ends of branches. What appear to be white flowers are dense white hairs in the cyathiums. Anthers and stigma are maroon red. Flowering occurs February to November.

Leaves are long and narrow. Upper leaves are in a whorl at the base of the flower cluster at the top of the plant, radiating out like the ribs of an umbrella. Lower leaves are alternate and very quickly deciduous.

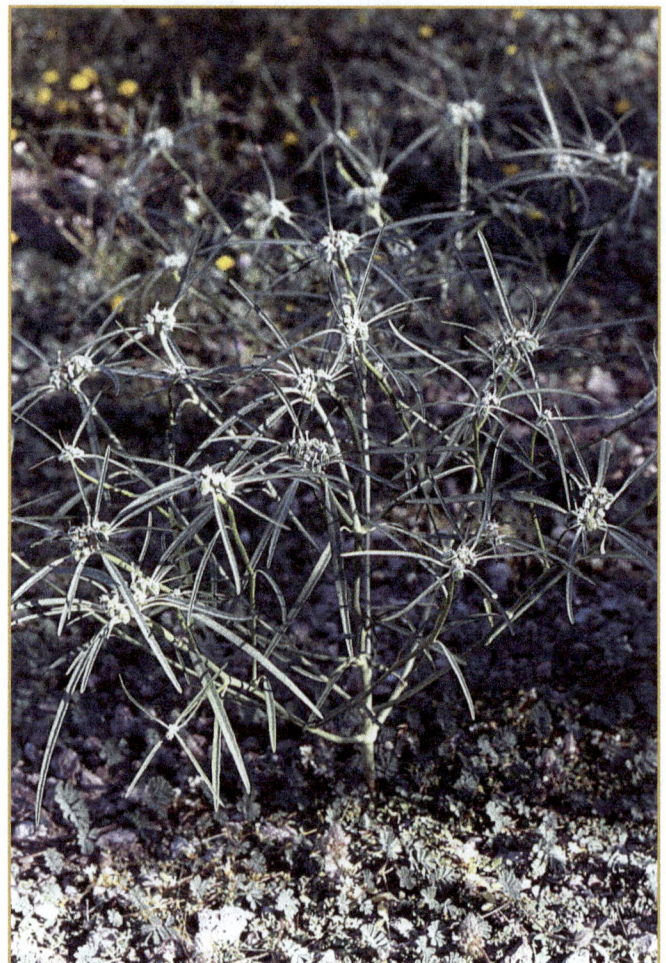

WHERE TO SEE IT
Beetle Spurge is common and widespread on sandy, gravelly, or rocky sites northwest of Kino Bay, and on stabilized inland sand dunes at Tastiota. It grows in response to rain and disappears during dry periods.

C-55 WHITE RATANY, Chacate
Scientific name: *Krameria grayi*
Family: Krameriaceae. *Ratany family*

White Ratany is a sprawling shrub to 3 feet high and several feet wide, with a dense tangle of very fine stems and branches. In our area, the stems are green on top and reddish underneath, turning black during extended dry periods. It is thornless, but branch tips are sharp. White Ratany is parasitic on the roots of Bursage and Creosotebush. It is browsed by cattle and wildlife.

Flowers are about ½ inch wide and very showy. Flower sepals are bright magenta-purple inside and

Alberto Mónica and a large White Ratany bush behind Kun-kaak R.V. at Kino Bay.

white hairy outside. The three upper petals are spear-shaped, bright chartreuse, with purple tips. The two lower petals are dark purple oil glands that are very thick and slab-like.

The fruit is round, about 1 cm across, with many short, thin spines, each having tiny barbs in a terminal whorl (lower right). Flowering occurs at various times. The main flowering period is April through September.

The stems are sparsely-leaved or often leafless. Leaves are 5 to 10 mm long and 1 to 2.4 mm wide, linear or oblong and often grayish.

White Ratany grows on dry desert slopes, arroyos, and hillsides. It is an occasional plant from northern Sonora, south to Kino Bay and Hermosillo, where it is replaced to the south by Sonora Ratany (*K. sonorae*).

Seri Indians make a rich, red-brown dye from the bark of the lateral roots of White Ratany. The bark is crushed and placed in boiling water. Basket splints are placed in the mix to allow the dye to be absorbed. This dye was used for the red-brown color on the dish-shaped basket pictured in C-22.

WHERE TO SEE IT
White Ratany grows in the desert flat behind Kunkaak R.V. and on sand dunes on vacant lots on the north side of the Mar de Cortés in New Kino. Also on sandy inland dunes just south of Puerto Libertád.

125

C-56 White Spurge, Jumetón, Liga,

Scientific name: *Euphorbia xanti*
Family: Euphorbiaceae. Spurge family

A shrub, 3 to 6 feet tall with thin, smooth, straight, greenish branches. It often grows under a host shrub or tree where the branches spread out on top of the host, giving the impression it is part of that plant. During dormancy, when the leaves and flowers are absent, the shrub is easily missed. The plant is related to Cliff Spurge.

The cyathium (see Cliff Spurge, E-3 for definition) has five petal-like appendages that are white or pink. The appendages are about 3 mm long. Cyathiums are numerous and very pretty. Flowering occurs from October through May, usually after rains.

Leaves are highly variable. Size ranges from 1 to 24 mm wide and 1.5 to 3.5 cm long. They may occur singly or in groups of three to six. They may be linear or rounded. Leaves are drought-deciduous.

White Spurge is locally common in washes, sandy flats, and rocky slopes. It occurs in the lower two-thirds of the Baja peninsula, but its mainland distribution is largely centered in a small area between Kino Bay and El Desemboque.

White Spurge is an ornamental in southern California and Arizona. It is easily grown from stem cuttings or seed. The milky sap is poisonous. The sap of many plants in the spurge family is either caustic or poisonous.

WHERE TO SEE IT
The pictures above are of White Spurge cyathiums (flowers) growing on top of a host shrub along the entry road to Dos Palmas residential area about 10 miles northwest of Kino Bay. White Spurge is rather common on rocky bajadas between Kino Bay and El Desemboque.

126

Area D

Tastiota Area South of Kino Bay

D-1 (No common name)
Scientific name: *Grusonia (Corynopuntia,Opuntia) reflexispina*
Family: Cactaceae. Cactus family

A low (4 to 16 inch high) cholla-like cactus growing in clusters up to 4 feet across. A very rare plant, limited to an area between the sea at Tastiota and Arrieros, Sonora. Grows on stabilized, inland dunes, with an occasional plant group growing on clay soils in the small blow-outs between these dunes. Some growing under the protection of a host shrub attain a height of 24 to 30 inches with groups 7 feet wide.

The large cup-shaped flowers are 1 to 1½ inches wide by 1 inch long. Flowering occurs from late July into September. Time of flowering appears to be weather-dependent. Flowers are light yellowish-green, appearing nearly white from a distance. Flowers are bright and attractive. In the experience of the author, flowers open gradually, beginning in late morning, and are fully open by mid-afternoon, closing at night.

The spines are about ½ inch long, growing in clusters along the stems. The upper spine within each cluster extends horizontally while the others point downward, thus the species name – "reflexispina," or reflexed spine.

The author found worms feeding on the inside of flower buds of several plants. Although this may be a detriment to reproduction by seed, it appears most, if not all, reproduction is vegetative.

G. reflexispina is similar and closely related to *Grusonia marenae*, another rare cactus found at widely-scattered locations from Kino Bay, north to Caborca, Sonora. Ranges of the two do not overlap. Both have large tuberous roots that were once eaten by the Seri Indians.

Although designated a "Species of Special Concern" in Mexico, nothing is being done to halt destruction of habitat from encroaching shrimp pond construction at Tastiota.

WHERE TO SEE IT
Vicinity of Tastiota, Sonora.

128

D-2 CHUPONES, Toje

Scientific name: *Psittacanthus (Phrygilanthus) sonorae*
Family: Loranthaceae. Mistletoe family

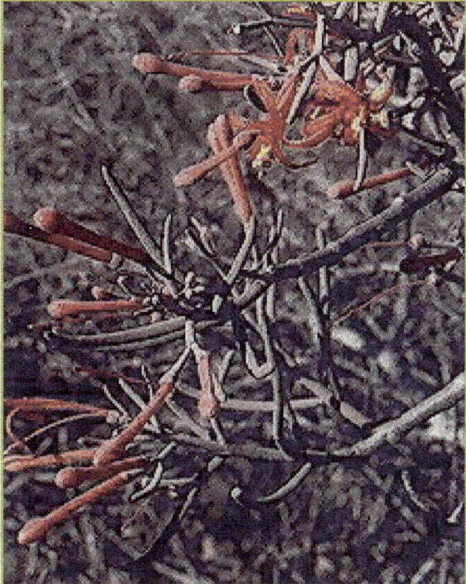

Chupones is a woody, parasitic mistletoe that infects elephant trees (*Bursera* spp.) in our area on the mainland. It has stout, stubby branches 4 to 16 inches long. The crimson red, tubular flowers appear October through December and give the impression they are the flowers of the host. The flowers have six red sepals, but no petals. Sepals are 1 to 1½ inches long and dark red. Flowers are pollinated by hummingbirds. The fruits are oblong, ½ inch long, smooth, purplish to nearly black.

The leaves are terete (round in cross section), 1½ to 3 inches long, and linear. Chupones is not common on the mainland.

The name "Toje" is a generic Mexican term for several species of mistletoe.

In Baja California, Chupones can be found growing on elephant trees from Bahía de Los Angeles, south to the Cape Region (mountain range extending 100 kilometers south from La Paz), and on several Gulf islands. It also grows on Wild Plumb (*Cyrtocarpa edulis*), an endemic in the Cape Region.

Chupones does little harm to Wild Plumb but appears to be quite harmful to elephant trees.

Chupones flowers and the plant's point of attachment to the host tree look quite natural, leading some to believe the host has two kinds of flowers (Norman C. Roberts, *Baja California Plant Field Guide*).

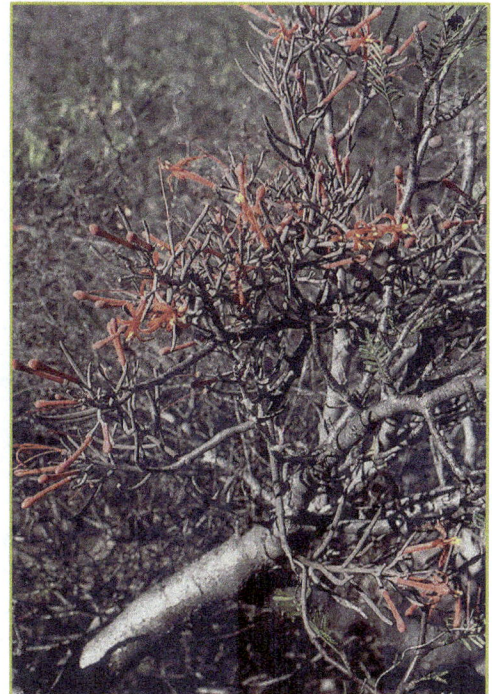

WHERE TO SEE IT
Pictures were taken on the road to El Colorado, near the coast just south of Tastiota, Sonora. Chupones had infected a group of Little-leaf Elephant Trees (*Bursera microphylla*) on the north side of the road a short distance from El Colorado. All trees appeared healthy. The author has not seen Chupones in the Kino Bay Area.

D-3 SLIPPER PLANT, Candelilla, Zapote del Diablo (Shoe of the devil)

Scientific name: *Euphorbia llomelii*
Family: Euphorbiaceae. Spurge family

A clump-forming succulent shrub with waxy stems to 40 inches tall. The stems are smooth, gray-green, and exude a white milky latex when injured. The small (to 1 inch long) leaves fall soon after emerging, leaving the plant bare most of the year. Root sprouting sometimes creates extensive colonies.

The most striking feature of the plant is the bright red slipper-shaped flower which is really a group of one female flower (which develops into a fruit) and several male flowers. This floral characteristic is typical of the spurge family. Flowers appear February through May and August through October. The generic name "*Pedilanthus*" is Greek for pedilon (slipper or sandal) and anthus (flower).

On the mainland, Slipper Plant is restricted to silty flats of the coastal plain in the states of Sonora and Sinaloa. The northern limit is reached at Tastiota, south of Kino Bay. It is more widely distributed in the southern two-thirds of the Baja peninsula but is seldom abundant.

At one time, Mexicans cooked the stems and made candles from the plant's waxy coating. In addition, the milky latex from inside the stem contains 6 to 10% good quality natural rubber, and wild stands were harvested for rubber in the early 1940s. The somewhat toxic milky sap is used by natives for chapped lips, cuts, and burns, but may cause diarrhea.

Slipper Plant is occasionally used as a landscape plant in Kino Bay and elsewhere in Sonora and the American Southwest. The author has found plants still thriving in flower beds of abandoned ranches in the desert near Kino Bay. Slipper Plant requires very little water and

130

can survive extended droughts. It is available in some nurseries in the American Southwest. It prefers well-drained soil and direct sunlight.

A close relative and look-alike which is also called "Candelilla" (*Euphorbia antisyphilitica*) is a native to parts of southwestern Texas, ranging to Hidalgo and Queretaro, Mexico. The stems are thinner and shorter than *P. macrocarpus*. At one time, it was common along the Rio Grande, where it was harvested for its waxy coating and used to make candles. Over-harvesting eliminated the plant from much of its native range; however, it is available in nurseries in the Southwest. The milky sap was once thought to be a cure for syphilis, thus the species name "*antisyphilitica*."

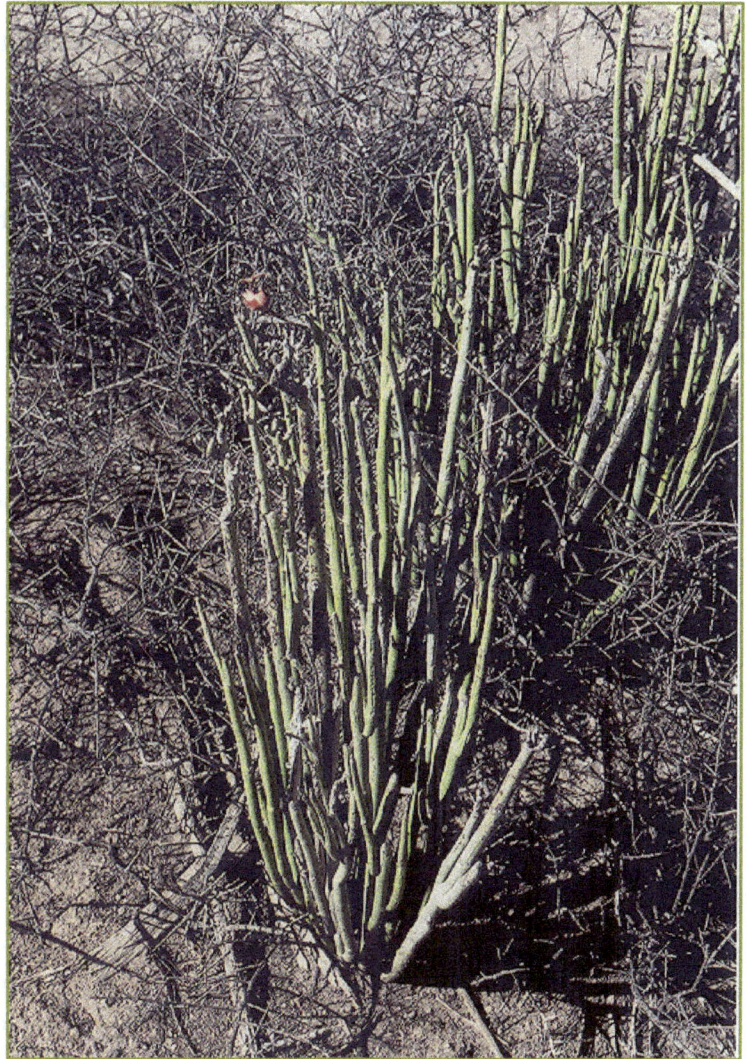

WHERE TO SEE IT
Slipper Plant can be seen growing on flats along the dirt road between Calle 4 Sur and El Colorado, south of Tastiota, Sonora. It grows within a few miles of the sea and is commonly found growing under or close to other shrubs. During flowering time, the bright red flowers are easily seen from a distance.

D-4 ADAM'S TREE, Palo Adan

Scientific name: *Fouquieria diguetii*
Family: Fouquieriaceae. Ocotillo family

Adam's Tree is one of four ocotillos (A-2, C-5, D-4, and E-1) covered in this book. It is common on the lower half of the Baja peninsula but has limited distribution on the mainland, where it is found only in the Tastiota to Guaymas area. In the Tastiota area, Adam's Tree is commonly a small tree, 10 to 20 feet tall, with a short, thick trunk and at least some of the stems much thicker than the others. Mature trees have a trunk to 8 inches in diameter, stem bark that is diagonally striped, and a crown spread up to 20 feet. Mature trees are quite attractive. Bark is red-brown.

The flowers are bright red, about 1 inch long, and grouped in large panicles. Unlike other species of ocotillo, Adam's Tree does not have a concentrated bloom period. It produces solitary inflorescences generally from February through May. The author has found some of the Tastiota trees with a few flowers as late as October following an especially wet summer.

Leaves are leathery and elliptic to ovate. Short shoot leaves are from ½ to 1 inch long. Long shoot leaves are ⅔ to 1¾ inch long. Leaves appear after rains and drop off during dry periods.

Refer to C-5, Common Ocotillo, for pictures of the floral difference of ocotillos and a description of short and long shoot leaves.

When young, most stems are about the same size and Adam's Tree closely resembles common Ocotillo. As the tree matures, some stems will have much larger diameters. Both ocotillos share the same sites at Tastiota. Adam's Tree is closely related to Tree Ocotillo (A-2) whose range overlaps Adam's Tree in the vicinity of Guaymas but is not known to occur at Tastiota. Adam's Tree is the common ocotillo over the southern two-thirds of the Baja peninsula. Its distribution on the mainland is limited to a small area from Tastiota to Guaymas and is thought to be a disjunct population originating from the peninsula.

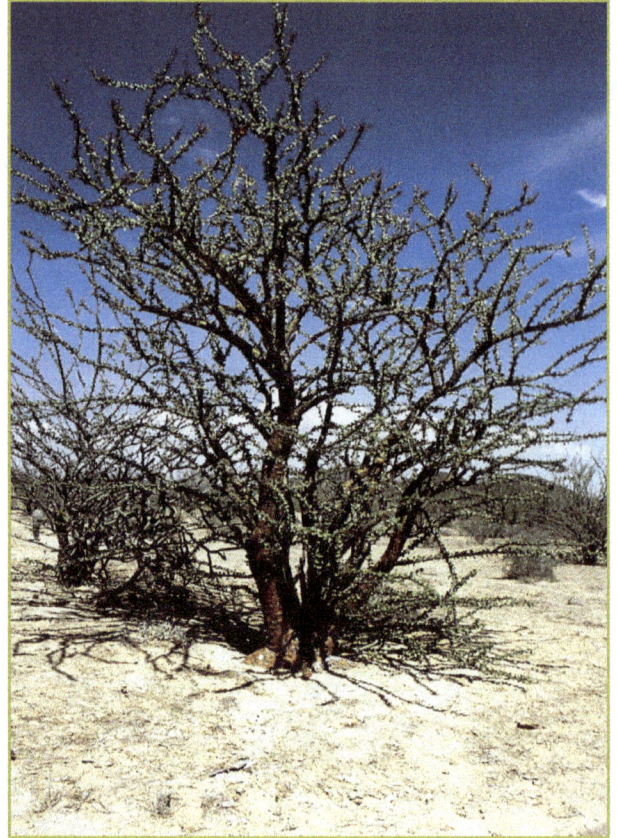

WHERE TO SEE IT

Adam's Tree is growing on a flat at the foot of a hill on the east side of Calle 4 Sur opposite the junction to Tastiota. Drive south past this junction about ¼ mile, turn left on a primitive dirt road and go through a gate. There are many Adam's Trees visible in the flat after entering through the gate.

132

D-5 STAGHORN CHOLLA

Scientific name: *Cylindropuntia (Opuntia) versicolor*
Family: Cactaceae. Cactus family

A tree-like cholla 5 to 7 feet tall in our area but may reach 16 feet tall in other parts of its range. It has a large, open-branching top up to 10 feet wide. Most have a distinct, single, woody trunk, 3 to 4 inches in diameter and up to 3 feet long. Staghorn Cholla resembles Buckhorn Cholla (*Cylindropuntia acanthocarpa*) and often grows on the same sites.

Stems are green or tinged with purple (photo below), about 1 inch in diameter, with spine clusters of six to twelve gray to purplish spines, ⅝ inch long (longer spines to 1 inch long). Nearly all spines in the spine cluster are of equal length.

The 1 to 2 inch wide flowers may be in any of six colors; yellow, orange, greenish, red, brown, or bronze. Flower colors are uniform on any one plant but may vary strikingly in a colony. According to Richard Felger, all Staghorn Chollas south of Hermosillo have yellow-green flowers (*The Trees of Sonora Mexico*, 2001. by Richard Felger. Oxford University Press). Flowering occurs from March through May.

The fruit is bright yellow, fleshy, pear-shaped, 1 to 1½ inches long, about 1 inch in diameter, with an abrupt excavation at the tip 3 to 6 mm wide and as deep. The fruit has glochids, but rarely spines. The fruit usually stays on the plant the entire year and sometimes develops chains of two to four fruits.

In our area, Staghorn Cholla becomes more common in and around foothills and mountains and less common in the drier open desert. It is also found in arroyos, washes, and at the mouths of canyons, showing a preference for more moist habitats. Its range in Sonora is from northern to central Sonora and from sea level to 3,000 feet in elevation.

WHERE TO SEE IT

A signed Staghorn Cholla is in the cactus garden at the Highway 15 toll booth at Magdalena, Sonora. Also, an excellent population can be seen near Tastiota. From the junction of Calle 4 Sur with the road west to Tastiota, drive south about ¼ mile. Turn left (east) on a primitive dirt road and go through a gate. There are many Staghorns on this flat.

D-6 LOLLIPOP TREE, Palo San Juan, Jito, Palo Jito

Scientific name: *Forchhammeria watsonii*
Family: Capparaceae (Capparidaceae). Caper family.

A thornless tree, to 35 feet tall, with a dense, rounded, or scraggly crown. The trunk is pale gray and fairly smooth. Some plants become trees, while others are sprawling, dense shrubs caused by heavy livestock grazing.

The tree has both male and female flowers that are very tiny. There are sepals but no petals. Male flowers are yellow. Female flowers are maroon. The fruit is ovoid, about ½ inch wide, purplish to reddish-orange, with a sweet pulp. The fruit is boiled and eaten with or without sugar. Flowering occurs March and April with fruit produced May through June and sometimes November.

Leaves are long and narrow (1½ to 4 inches long and 8 to 10 mm wide), stiff, narrow, dark green above, pale green beneath, with edges rolled inward. Leaf margins are smooth. The tree is evergreen except for a brief period at flowering time.

Lollipop Tree can spread by lateral roots near the soil surface to form additional trees. It makes its best growth on deep alluvial soils; however, most trees seen along Calle 4 between Tastiota and Highway 15 are growing on rocky slopes and, more commonly, in rocky arroyos.

It is heavily browsed by livestock, resulting in the rounded (lollipop) canopy of shrub-sized trees.

HOW TO TELL THEM APART:

The tree closely resembles Pinicua (*Jacquinia macrocarpa*). See C-6. Leaves of Lollipop Trees are unarmed (no prickle at the leaf tip). Leaf tips of Pinicua have a sharp prickle. Both species occur in this area.

WHERE TO SEE IT

Lollipop Trees can be seen growing along Calle 4 Sur about half way between Tastiota and the junction with Mexican Highway 15. Some are growing in the fence on the north side of the highway. The picture was taken of a tree next to a parking area behind some buildings at this junction.

Area E

Puerto Libertád Area
North of Kino Bay

E-1 BOOJUM, Cirio

Scientific name: *Fouquieria (Idria) columnaris*
Family: Fouquieriaceae. Ocotillo family

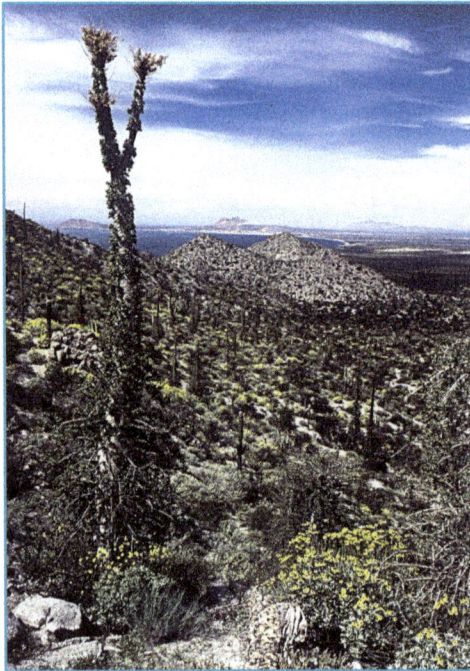

One of the most bizarre plants of the Sonoran Desert is a carrot-shaped tree called the Boojum. Although common in the central third of the Baja peninsula, the only mainland population occurs in a narrow strip 30 miles long and 3 miles wide along the coast from Puerto Libertád, south to Desemboque. Here, they are most common on north slopes with granitic soils of the Sierra Cirio.

Godfrey Sykes, a scientist with the former Carnegie Desert Research Laboratory at Tucson, is credited with the name, which comes from a mythical creature in the Lewis Carrol book, *The Hunting of the Snark*. When Sykes first saw the tree in 1922, he said: "A boojum, definitely a boojum." Sykes Crater in the Pinacate volcanic field near Puerto Peñasco, Sonora is named for him.

Jesuit padres who first saw the tree in 1762 called it "Cirio" (see-ree-oh), which is a slender alter candle used in Mexican churches. Mexicans know it by that name.

The Boojum is one of four species of ocotillo in Sonora. It produces leaves after seasonal rains and sheds them in dry periods. It is usually leafless and dormant in summer. Pale yellow or white flowers appear on stick-like stalks at the top of the tree any time after rainfall but commonly July through September. Boojums have an extensive root system that allows them to take up and store a large amount of water during the infrequent rains. They have a single tapering trunk that occasionally divides into two or more branches near the top. The bark is thick and corky. Underneath are a woody cylinder and a large central core of pith for water storage. Unlike Saguaro, the woody cylinder does not expand with water uptake. Consequently the tops of some become so heavy with water they may bend down and eventually grow into a "U" or even form loops. Unlike the other ocotillos, the main stem grows only in winter following sufficient rain. There may be no growth during some years.

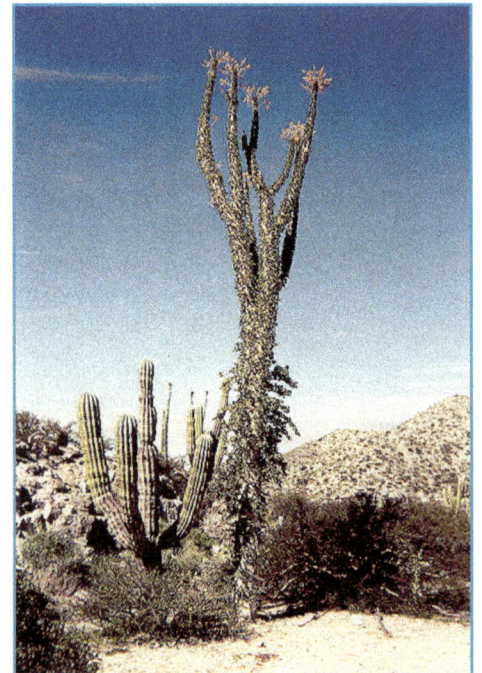

There are millions of Boojums in the central third of the Baja peninsula, from El Rosario in the north to the Tres Virgenes Mountains near Santa Rosalia in the south. There is a small population at the northern tip of Angel de la Guarda Island off the east coast of central Baja. Boojums on Baja grow best on rocky granitic hills and alluvial plains where winter rains are dependable and the desert climate is moderated by the cool, moist winds from the Pacific Ocean. Few grow east of the peninsular mountain crest. Total annual precipitation for the Boojum habitats on Baja and the mainland is about 5 inches.

Boojums on the mainland reach a height of 30 to 40 feet. The tallest measured 52 feet. Their counterparts on the Baja peninsula can be taller, but most seen along the peninsular highway are only 20 to 30 feet tall. While doing research on Boojums in 1970, Robert Humphrey discovered the tallest one in Montevideo Canyon near Bahía de Los Angeles on the east side of Baja. It measured 81 feet and grew several more feet in the next twenty years. This monster and a 60-foot tall Cardón Cactus growing beside it were both gone when the site was revisited in 1998, probably casualties of Hurricane Nora the previous year.

There is no accurate way to measure the age of Boojums. Humphrey thought they might reach an age of 700 years; however, recent studies show the life span to be only about 100 years or more. Photos taken in a 1905 study showed many tall Boojums and Cardón Cactus. Photos retaken 90 years later showed none of the tall trees had survived. Boojums, Cardón Cactus, and other species of tall columnar cactus are vulnerable to high winds and suffer significant losses from hurricanes.

There has always been the question of why there is an isolated population of Boojums on the mainland. Are they a remnant of a former population when the Baja peninsula split off from the mainland beginning 150 million years ago, or just a chance occurrence from seed dispersal? The habitat requirements on the mainland site fit those in central Baja, except for the cool, moist winds of the Pacific. Some scientists think there may be an upwelling of cool water in the Sea of Cortés adjacent to the mainland population that provides this condition.

In Seri mythology, it is believed the first Seri were giants. In order to escape a great flood, a group of giants from the south fled north to the mountains between Desemboque and Puerto Libertád. There the flood overtook them and they changed into Boojum trees. The Seri believe these trees have much power from the spirit "Icor" and if touched or harmed, will cause strong winds or rains.

Young plants are easily transplanted, and Boojums can be grown from seed. Several nurseries in Tucson sell Boojums. There are a number of transplants on the campus of the University of Arizona in Tucson.

WHERE TO SEE IT

There is a transplant on a vacant lot in New Kino on the north side of Mar de Cortés just west of Calle Palermo (photo above). The mainland population can be seen on the hills just south of Puerto Libertád. There are no signs. Look for a dirt road going west from Calle 36 Norte near where the power line crosses the highway. The dirt road forks immediately after leaving Calle 36. Take the left fork and follow it southwest to a locked gate. Boojums are visible in the hills to the south. They are on private land, but it is permissible to go through the fence and walk to the trees.

E-2 TEDDY BEAR CHOLLA, Jumping Cholla, Cholla Güera

Scientific name: *Cylindropuntia (Opuntia) bigelovii*
Family: Cactaceae. Cactus family

An upright cholla, 3 to 6 feet tall, with husky stems and usually only one trunk which turns black with age. It has dense, short side branches in the top one foot, and long (1 inch), very dense, golden spines. The spines are in groups of seven to twelve. Teddy Bear Cholla usually forms extensive stands on rocky or sandy ground over much of its range. In the Kino and Puerto Libertád areas, it usually grows in small groups or singles among other species of cholla.

Flowers appear in clusters at the ends of joints from February to May. Flowers may be greenish-yellow, pale green, or white streaked with pale lavender. Fruits are often sterile. The plant reproduces primarily from stem segments called "joints" that break off and take root

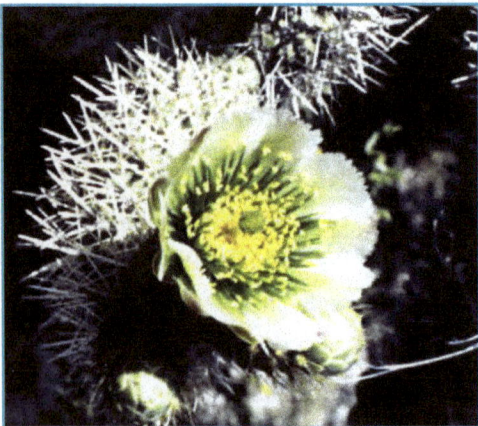

at the base of the parent, or at some distance after being carried by water or animals.

Teddy Bear Cholla most closely resembles Chainfruit Cholla which also grows in the same area.

HOW TO TELL THEM APART:

Both varieties of Chainfruit Cholla have long chains of fruit. There are no chains on Teddy Bear Cholla. Joints on Chainfruit Cholla tend to droop, giving the plant a scraggly appearance. Joints on Teddy Bear Cholla are thick, have dense yellow spines giving them a furry appearance, and the joints tend to stick straight up at the top of the plant.

Teddy Bear Cholla inhabits the drier parts of the Sonoran Desert. The dense armament of spines is thought to have developed as a protection against animals and, to a lesser extent, for protection from the intense heat in its range. Packrats (Neotoma spp.) gather joints of this and other chollas to pile on top of their stick mounds for protection.

Yellow fruit (tunas) and dense golden spines on joints.

Teddy Bear Cholla grows on rocky, south- or west-facing slopes, and gravelly flats and rolling hills. It is uncommon in the immediate area around Kino Bay but can be found growing in small patches on high ridges in the mountains just northwest of Kino. It is common on the inland dunes directly south of Puerto Libertád.

Joints from the current year's growth are edible. Break off the joints, burn off the spines and place in a covered pit to cook for 30 minutes. Remove and eat.

The general view photo of a Teddy Bear Cholla (previous page) was taken in Joshua Tree Parkway near Kingman, Arizona, which is not in the Sonoran Desert. Note the Joshua Tree in the rear and a pink-flowered Beavertail Cactus in the foreground. The Joshua Tree is the symbol species of the Mojave Desert. Its distribution only touches the Sonoran Desert near Needles, California. The Beavertail Cactus occurs in both Great Deserts.

CAUTION: The joints break off from the slightest touch, and the spines quickly imbed in the unfortunate animal or person. If this should happen, do not try to shake or brush off the joint as it further imbeds the spines. A comb works best to brush off the joint. Remove any remaining spines with pliers or tweezers. A rash or darkening of the affected area may occur, caused by a sudden rush of blood to the skin surface. Contrary to popular belief, the spines do not contain a poisonous property.

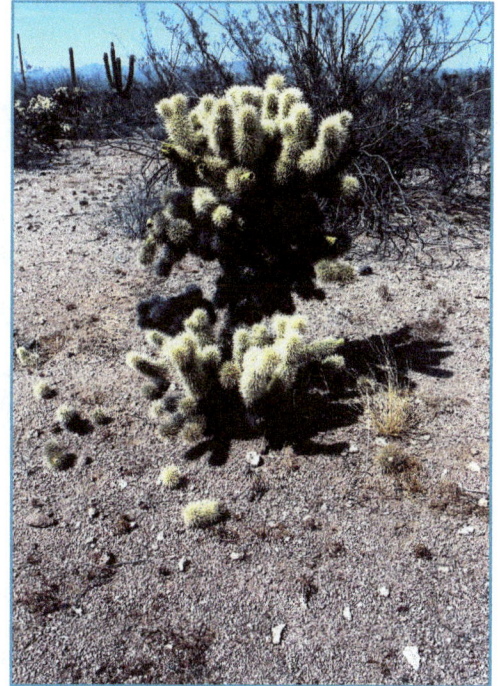
New plants growing from joints at the base of a parent plant. Other joints may be carried some distance by wind and water to begin new colonies. Brown spots on the ground are dead joints.

WHERE TO SEE IT
Inland dunes just south of Puerto Libertád. The golden spines make this cholla easy to spot.

E-3 CLIFF SPURGE, Jumetón

Scientific name: *Euphorbia misera*
Family: Euphorbiaceae. Spurge family

A straggling, rounded shrub, 2 to 5 feet tall, with thick, mahogany-colored, branches. The stems are thick, semi succulent, knotty and flexible, and contain a white milky sap. The knotty, flexible stems somewhat resemble Wedgeleaf Limberbush, which is also a member of the spurge family.

The flower in all spurges is called a cyathium which consists of male flowers, a female flower, and glands with petal-like appendages. For Cliff Spurge, the cyathium is 3 to 12 mm wide. In the photo below, what appears to be a petal is a purplish gland with a white, petal-like appendage.

The fruit is a globose, three-lobed capsule, 1½ to 2 inches long. Flowering occurs sporadically throughout the year following rains. The cyathiums (flowers) are very attractive.

Leaves are pale green, orbicular or oblong, ¼ to ½ inch long and mostly on short spur shoots. They appear shortly after winter or summer rains and drop off when the soil dries. The short spur shoots give the stems their knotty appearance.

On the mainland, Cliff Spurge occurs sporadically along the coast from about Desemboque (south of Puerto Libertád), north to Rocky Point. It is also found in the Pinacate Mountain range and on Tiburón Island. It is much more common along the Pacific coast of the Baja Peninsula.

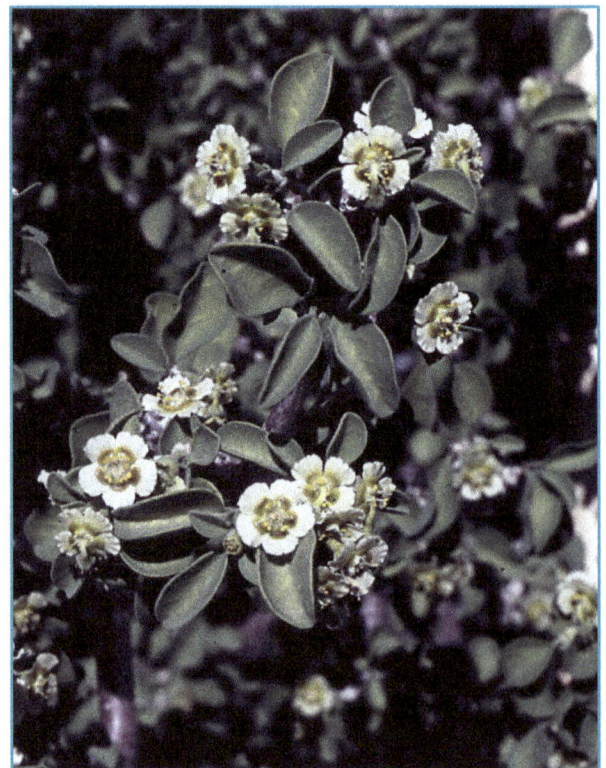

WHERE TO SEE IT
Cliff Spurge is rather common on inland dunes near the coast just south of Puerto Libertád. The picture (above) was taken looking north toward Puerto Libertád with the Sea of Cortés in the background at upper left.

E-4 MARIOLA, Sonoran Nightshade, Ojo de Liebre, Tomatillo Espinoso

Scientific name: *Solanum hindsianum*
Family: Solanaceae. Nightshade or potato family

An open-branched shrub, usually about 3 to 4 feet tall, but can reach 9 feet. The branches are gray to mottled and have spines to ½ inch long. Eye-catching when the lavender-purple flowers are in bloom.

The leaves are 1 to 2 inches long and ¼ to ¾ inch wide, obovate to oblong, and densely hairy. The leaves are evergreen except during periods of prolonged drought.

Flowers are lavender to purple, 1 to 2 inches wide and single. Berries are 1 to 2 cm wide, pale green with dark green stripes like a watermelon. Flowers and fruits are produced anytime of year following rain.

Mariola prefers well-drained rocky or gravelly sites along arroyos and on hillsides and bajadas.

Mariola is common over much of the Baja peninsula but is only an occasional plant on the mainland in Sonora. It can be found along the coast from Desemboque, north to the Pinacate Mountains and Sonoyta, Sonora. It also grows near the coast between Tastiota and Guaymas and on Tiburón and the other midriff islands. It does not grow at Kino Bay.

WHERE TO SEE IT
Mariola can be found near the coast south of Puerto Libertád growing on stabilized dunes and decomposed granite soils in the vicinity of the Boojum trees.

141

Glossary

Allelopathy – ability of some plants to produce a toxin to discourage establishment of other plants. See Buffelgrass (A-6).

Anther – pollen-bearing part of the stamen in a flower.

Annual – completing a life cycle in one growing season.

Areole – when the term applies to a cactus, it is a tan or gray-colored spot on the surface of a cactus that subtends the spine or spine cluster. See photo in Cardón cactus (C-10).

Bajada – a sloping, fan-shaped outwash of rocks and gravel at the base of a mountain. Not the same as stream terrace which occurs further out from the mountain.

Bosque – a forest of any woody plant. Example: Mesquite bosque.

Bristle spine – thin, bristle-like spine. See Arizona Barrel Cactus (C-14).

Calyx – outer series of petal-like parts of a flower. Inner series (petals) is called the corolla.

Catkin – a dense spike of flowers such as in willow.

Cardónal – a forest of Cardón cactus (C-10).

Central spine – usually the longest spine(s) in the center of a spine cluster. See also radial spine and bristle spine.

Cladophyll - (also called "phylodic petioles") - leaf-bearing petiole that manufactures chlorophyll like a leaf. Rare in the plant world. See White-bark Acacia (C-4).

Compound leaf – a leaf divided into two or more leaflets such as Western Honey Mesquite (C-20). Compare with "simple leaf."

Columnar cactus – Any tall cacti whose upright growth form resembles a column.

Convergent evolution – plants or animals developing similar characteristics in different geographic areas.

Corolla – a set of flower parts composed of the petals.

Cristate (Cresting) – grotesque growth form occasionally occurring in some species of plants. See Saguaro (C-9).

Cultivar – cultivated variety.

Cuneate – wedge-shaped or triangular with the narrow point at the place of attachment such as in a cuneate leaf or the wedge-shaped dent on a leaf tip.

Cyathium – The ultimate inflorescence in the genus *Euphorbia* consisting of unisexual flowers congested within a cup-shaped involucre.

Disk flower – a small, tubular flower in the center of the flower head in the sunflower family. The center of a Brittlebush flower is made up of numerous, tiny disk flowers.

Ecology – the study of the interrelationships among plants, animals and their physical environments.

Ecosystem – The combined living organisms and non-living physical components (i.e. soils, water, weather, etc.) within a distinct environmental unit and their relationships within this unit. (If you liked this, you'll love Mother Goose!).

Endemic – growing or living exclusively within a particular region or locality. See White-bark Acacia (C-4).

Exfoliate – to peel off in layers. See Little-leaf Elephant Tree (C-24).

Facultative – Capacity to live under more than one set of environmental conditions.

Filament – When applied to flowers it is the stalk of a stamen supporting the anther.

Fire-intolerant – plants that have evolved in the absence or near-absence of wild fire which are easily damaged or killed from fire. Example: cacti.

Forb – See herb.

Foreshaft, Arrow foreshaft – short stick, usually about 8" long, bearing the arrowhead and inserted into the main arrow shaft. Also used on harpoons.

Genus – scientific term used in biological classification.

Glabrous – smooth, without hairs or glands.

Glochid – a minute, barbed hair or bristle (minute spine). See Santa Rita Prickly Pear (A-13).

Glomerule – a densely crowded cluster, usually of flowers. The male flowers of plants in the genus *Euphorbia* occur in "glomerules".

Half shrub – a perennial plant having a main stem with a woody base and herbaceous (non-woody) stems above.

Habitat (Plant habitat) – an association of plant species that occur on a unit of land due to a specific set of soil and other environmental conditions.

Halophyte – Plant adapted to growth in salty soil.

Herb – a plant with no persistent woody stem above ground. Same as a forb. Compare to a shrub, tree or grass.

Inflorescence – the flowering part of the plant. The inflorescence may have (1) a single flower on a stem, (2) branches bearing flowers as in a "raceme," or (3) a single stalk bearing numerous flowers as in a "spike." See raceme and spike.

Involucre – A whorl of distinct or united leaves or bracts just below a flower or the stalk of an inflorescence.

Keystone species – a plant whose presence favors, or is essential for, the establishment of other plants and associated biota. See Western Honey Mesquite (C-20).

Leaf succulent – plants that store water in the leaves such as *Agave*. By contrast, *Yucca* is a "semi-succulent-leaved" plant and Saguaro is a "stem succulent" plant.

Maritime influence – weather conditions created or influenced by the presence of a large body of water. These conditions favor establishment of certain species of plants. See Cardón cactus (C-10).

Mesic – (mesic site, mesic plant) - Requiring a moderate quantity of water. Term can apply to a plant or its habitat. Compare with xeric.

Native desert – area supporting largely native desert plants not materially altered by man.

Naturalized plant – a plant that is foreign to a site but established there and reproducing as if native. Example: Buffelgrass (A-6).

Nocturnal anthesis – flowering at night. See Sonoran Queen of the Night (C-19).

Opportunistic – ability of some species of plants to quickly establish on a site. Term usually used when describing certain invasive, non-native plants. See Buffelgrass (A-6).

Passive diffusion – a process whereby some plants can take up water only when the soil is more moist than the plant's interior. See Saguaro (C-9).

Pedicel – the stalk of a flower.

Perennial – a plant normally living for three or more years.

Pinna – the first primary division of a pinnately compound leaf. Includes the rachis with its attached leaflets. Pinna is singular. Pinnae (plural) consists of two or more pinna. Example: The leaflets of Mexican Palo Verde (A-7) grow on the rachis of a very long pinna.

Pinnate – a leaf that is divided into leaflets or lobes along each side of a rachis. Mexican Palo Verde (A-7) has a pinnate leaf.

Pistil – female part of a flower consisting of the ovary, style and stigma.

Plant aspect – a visual impression of the two or more plants that appear most obvious on a natural (undisturbed) site. Not necessarily the most abundant species. Example: Mesquite-Palo Verde plant aspect.

Plant diversity – variety of different plant species.

Plant frequency – relative abundance of all plants in a given area.

Plant key – book used to identify plants scientifically based on contrasting characteristics.

Playa – As used by geologists, the term means the bed of a shallow lake with no outlet that holds water seasonally. It is usually barren and has soils of high clay content. Playas are common along the highway entering Kino Bay. Another definition is a sandy beach such as Playa San Nicholas just south of Kino Bay.

Puberelent – minutely hairy with soft, fine, short hairs.

Pubescent – bearing hairs of any sort.

Raceme – an inflorescence with pedicelled (stalked) flowers borne along a more or less elongated axis with the younger flowers near the apex. See Tree Ocotillo (A-2). Compare to "spike".

Rachis – the main axis (1) to an inflorescence or (2) of a compound leaf.

Radial spines – a group (whorl) of spines in a spine cluster, usually intermediate in size between central spines and bristle spines. See picture of spine cluster for Sonora Barrel (C-15).

Ray flower – flower in the sunflower family with a single strap-shaped corolla resembling a petal. Example: Brittlebush (C-28) flower is composed of an outer ring of ray flowers and a center made up of disk flowers.

Rhizome – underground stem producing roots and shoots at the nodes.

Root crown – the uppermost portion of the root that gives rise to the aerial stem(s) of the plant. Usually a thickened area close to the ground surface.

Sarcocaulescent – A plant with fleshy stems as in Red Elephant Tree (C-25) and Limberbush (C-22).

Semi-succulent – A plant with somewhat succulent leaves such as *Yucca*. Compare to leaf succulent plant such as *Agave*.

Simple leaf – leaf with a single leaf-like blade. The blade may be toothed, lobed or smooth along the edges but NOT divided as in a compound leaf. Example: Brittlebush (C-28). Compare with compound leaf consisting of leaflets, as in Mesquite.

Sintales – grove of Senita (C-11) cactus.

Spike – an elongated, unbranched inflorescence consisting of a cluster of flowers that are stalkless, or nearly so. Example: Kidneywood (A-16). Compare to raceme.

Spine cluster – group of spines arising from a single point such as the "central," "radial," and "bristle" spines of Arizona Barrel Cactus (C-14).

Stamen – male part of flower consisting of the anther and the filament.

Stem succulent – a plant that stores water in the stem such as Saguaro.

Stellate – star-like. Often used to describe a pattern of minute hairs on a plant part that radiate out from a common center.

Stigma – That part of the pistil that receives the pollen. Usually at the top of the pistil.

Stoloniferous – a trailing shoot above ground rooting at the nodes.

Style – the stalk-like part of the pistil connecting the ovary and the stigma.

Thermal inertia – A process whereby some plants absorb heat during the day and release it at night such as Saguaro cactus (C-9).

Thornscrub – Term used to describe an area having trees and shrubs that are mostly drought-deciduous, often thorny, pinnate-leaved, and multi-trunked. This is the vegetation covering much of southern and southeastern Sonora and Sinaloa.

Tomentose – dense, wool-like covering of tangled hairs.

Trioecious – species having three variations of gender each on separate plants. An example is Cardón cactus (C-10), where some plants have only female flowers, some only male flowers, and some have both male and female flowers. Trioecious species are uncommon.

Vertical diversity – variety or range of plant heights in a given area. Often cited as a measure of wildlife habitat quality.

Xeric – having very little moisture; tolerating or adapting to dry conditions. Compare with "mesic".

Xerophyte – a plant adapted to dry conditions.

Selected References

1. *Arizona Flora.* Thomas H. Kearney and Robert H. Peebles
2. *Baja California Plant Field Guide.* Norman Roberts*
3. *The Boojum and Its Home.* Robert R. Humphrey.
4. *Discovering the Desert.* William G. McGinnies
5. *Cacti.* Frank D. Venning.
6. *Flora of Baja California.* Ira L. Wiggins
7. *Cactaceas de Sonora, México: Su Diversidad, Uso y Conservación.* Rafaela P. Aguilar, Thomas Van Devender and Richard Stephen Felger.
8. *Flora of the Gran Desierto and Rio Colorado of Northwestern Mexico.* Richard Stephen Felger.
9. *Handbook of Mexican Roadside Flora.* Charles and Patricia Mason.**
10. *Native Plants for Southwestern Landscapes.* Judy Mielke.
11. *Natural History of the Sonoran Desert.* Arizona-Sonora Desert Museum, Tucson, Arizona.***
12. *People of the Desert and Sea: Ethnobotany of the Seri Indians.* Richard Stephen Felger and Mary Beck Moser.****
13. *Sonoran Desert: The Story Behind the Scene.* Christopher Helms.
14. *Sonoran Desert Plants: An Ecological Atlas.* Raymond M. Turner, Janice E. Bowers, Tony L. Burgess.
15. *Plants of Arizona.* Ann Orth Epple.***
16. *Seventy Common Cacti of the Southwest.* Pierre C. Fischer, Janice E. Bowers and Tony L Burgess.
17. *The Trees of Sonora, Mexico.* Richard Stephen Felger, Mathew Brian Johnson and Michael Francis Wilson.
18. *Vegetation and Flora of the Sonoran Desert.* Forrest Shreve and Ira L. Wiggins. (Two volumes)

SUGGESTED READING:

* Pictures and descriptions of common plants in Baja California and Sonora.

** Drawings and descriptions of plants Mexico-wide.

*** Pictures and descriptions of plants in Arizona and Sonora.

**** Use of plants in the Kino Bay area by Seri Indians. Pictures and drawings. An excellent ethnobotany.

Relative Abundance of the Plants by Area

Key:
- **A:** Abundant
- **C:** Common
- **O:** Occasional
- **S:** Scattered
- **R:** Rare

		Area A	Area B	Area C	Area D	Area E
Acanthaceae						
C-27	*Justicia californica*	O	C	C	O	O
Aizoaceae						
C-7	*Mesembryanthemum crystallinum*		O	O		
Amaranthaceae						
B-6	*Salicornia bigelovii*		C			
Asteraceae						
C-51	*Bebbia juncea*	O	O	C	O	O
C-28	*Encelia farinosa*	C		C	C	C
C-40	*Eupatorium sagittatum*			S		
C-32	*Pectis papposa*		S	A		
C-42	*Perityle emoryi*	O		O	O	O
C-1	*Trixis californica*	O		O	O	O
Boraginaceae						
C-33	*Cordia parviflora*			C	O	O
Burseraceae						
C-25	*Bursera hindsiana*			C	O	
C-24	*Bursera microphylla*	O		C	C	
Cactaceae						
C-9	*Carnegiea gigantea*	C		C	C	O
D-1	*Grusonia reflexispina*				C	
E-2	*Cylindropuntia bigelovii*			O		C
C-17	*Cylindropuntia fulgida v. mamillata*	C		A	C	C
C-18	*Cylindropuntia leptocaulis*	O		O	O	O
D-5	*Cylindropuntia versicolor*	O		O	C	
C-15	*Ferocactus emoryi*			O	O	O
C-11	*Pachycereus (lophocereus) schottii*	S		O	O	O
C-11	*Lophocereus schottii*	C		A	C	C
C-16	*Mammillaria grahamii*	O		O	O	O
A-13	*Opuntia santa-rita*	S			S	
C-10	*Pachycereus pringlei*		S	C	C	C
C-19	*Peniocereus striatus*			O		
A-5	*Stenocereus alamosensis*	S			S	
C-12	*Stenocereus gummosus*			R		
C-13	*Stenocereus thurberi*	C		C	C	C

Capparaceae						
D-6	*Forchhammeria watsonii*				S	
Chenopodiaceae						
B-7	*Allenrolfea occidentalis*		A	O		
Convolvulaceae						
A-1	*Ipomoea arborescens*	O				
A-20	*Ipomoea ternifolia var. leptotoma*	S				
Euphorbiaceae						
B-4	*Croton californicus*		C			
C-54	*Euphorbia eriantha*			S	S	S
E-3	*Euphorbia misera*					O
B-8	*Euphorbia* spp.	C	C	C	C	C
C-56	*Euphorbia xantii*			O		R
C-23	*Jatropha cinerea*			C	C	C
C-22	*Jatropha cuneata*			C	C	C
D-3	*Euphorbia llomelii*			O		
A-3	*Ricinus communis*	S		S	S	S
Fabaceae (Leguminosae)						
A-17	*Acacia angustissima*	C				
C-34	*Vachellia (Acacia) farnesiana*	O		O	O	
C-4	*Acacia willardiana*			S		
A-12	*Calliandra eriophylla*	O				
A-19	*Coursetia glandulosa*	O		R	R	
A-16	*Eysenhardtia polystachya*	O				
A-11	*Mimosa dysocarpa*	C				
C-21	*Olneya tesota*	C		C	O	S
A-7	*Parkinsonia aculeata*	C		C	S	
A-8	*Parkinsonia microphylla*	C		C	O	O
C-36	*Parkinsonia praecox*	O		O	O	O
C-45	*Phaseolus filiformis*			S		
C-20	*Prosopis glandulosa*	A	S	A	C	C
Fouquieriaceae						
E-1	*Fouquieria columnaris*					A
D-4	*Fouquieria diguetii*				C	
A-2	*Fouquieria macdougalii*	O			R	
C-5	*Fouquieria splendens*	C		C	C	C
Frankeniaceae						
B-3	*Frankenia palmeri*		C	R		R
Hydrophyllaceae						
C-31	*Nama demissum*	S		S	S	S
Koeberliniaceae						
C-46	*Koeberlinia spinosa*	S		O	O	O
Krameriaceae						
C-55	*Krameria grayi*	O		O		O

Lamiaceae					
C-39 *Hyptis albida*	R	O	O	O	O
Loasaceae					
C-41 *Mentzelia adhaerens*	O		O	O	O
Loranthaceae					
D-2 *Phrygilanthus sonorae*				R	
Malpighiaceae					
A-15 *Callaeum macroptera*	C		O		O
Malvaceae					
C-49 *Abutilon palmeri*	S		S	S	S
A-18 *Gossypium thurberi*	O				
C-29 *Hibiscus denudatus*	C		C	C	C
C-48 *Horsfordia alata*			S		
C-30 *Sphaeralcea coulteri*	C		C	C	C
C-26 *Melochia tomentosa*			C	C	
Martyniaceae					
C-38 *Proboscidea althaeafoila*	S		S	S	S
Nyctaginaceae					
B-1 *Abronia maritima*		C			
Papavaraceae					
A-4 *Argemone platyceras*	C		O	O	O
Passifloraceae					
C-50 *Passiflora arida v. arida*	O		O	O	O
Poaceae					
A-6 *Pennisetum ciliare*	A		O	O	
Portulacaceae					
B-2 *Sesuvium portulacastrum*		C			
Rhamnaceae					
C-47 *Ziziphus obtusifolia*	C		C	O	O
C-53 *Colubrina viridis*	S		O		
Rhizophoraceae					
B-5 *Rhizophora mangle*		S			
Sapindaceae					
C-2 *Cardiospermum corindum*	O		C	O	O
Solanaceae					
C-3 *Lycium* spp.	C	S	C	C	C
C-37 *Nicotiana clevelandii*			O	O	O
C-35 *Datura discolor*	O		O	O	O
E-4 *Solanum hindsianum*				O	O
Stegnospermataceae					
C-52 *Stegnosperma halimifolium*			O	O	O
Theophrastaceae					
C-6 *Jacquinia macrocarpa var. pungens*			S	S	R

Zygophyllaceae					
C-43 Fagonia sp.			O		
A-9 Guaiacum coulteri	S		S	R	
C-44 Kallstroemia californica			O		
A-10 Kallstroemia grandiflora	C				
C-8 Larrea tridentata ssp tridentata	C		C	C	C

ABOUT THE AUTHOR

William J. Little has a degree in Forest-Range
Management from Colorado State University.
During a thirty year career with the U.S. Forest
Service, he worked in Oregon, Idaho, and Utah,
serving as either a forester, forest ranger, or
range specialist. After retiring from the Forest
Service, he was a range and environmental
consultant.

Bill has now turned his attention to one of the
more fascinating desert areas of the world and
has spent the past fifteen years roaming the
Sonoran Desert in Arizona, Sonora, and Baja
California to study and photograph its unique
variety of plants. The result is this guide to some
of the most interesting and unusual plants found
in the deserts lining the Sea of Cortés near Kino
Bay, Sonora, Mexico.

www.ingramcontent.com/pod-product-compliance
Lightning Source LLC
Chambersburg PA
CBHW052112020426
42335CB00021B/2735